D1046774

KING
ARTHUR
& THE GRAIL

Also by Richard Cavendish

A History of Magic

KING ARTHUR & THE GRAIL

The Arthurian Legends and their Meaning

RICHARD CAVENDISH

TAPLINGER PUBLISHING COMPANY
New York

Published in the United States in 1979 by
TAPLINGER PUBLISHING CO., INC.
New York, New York

LIBRARY OF CONGRESS CATALOGING IN PUBLICATION DATA

Cavendish, Richard.
King Arthur & the Grail.

Bibliography: p.
Includes index.
1. Arthurian romances—History and criticism.

I. Title.
PN685.C3 1979 809′.933′51 79-14034
ISBN 0-8008-4464-5

Printed in the United States of America

Contents

List of Family Trees

Preface

One or two points need to be made briefly at the outset. A good deal has been written recently about the real Arthur, the man round whom the legends gathered. I have kept my treatment of him to a minimum and have concentrated on the legends themselves. This book is about the characters, themes and motifs of the Arthurian and Grail legends, viewed as tales of heroism, honour, the search for integrity and the conquest of death. It is here, I believe, that the lasting value and attraction of the Arthurian cycle is to be found.

The history of the legends begins with the real Arthur, about AD 500, and ends for my purposes with Malory in the fifteenth century. It is an indispensable framework, but I have dismissed as many of the minor details as possible to Appendix I, which contains notes on all the Arthurian writers, books and stories mentioned in the main text.

Generally, I have given the names of the characters their familiar modern spellings. Arthur's wife, for instance, appears in different works and languages as Guinevere, Gwenhwyfar, Guanhumara, Guenièvre, Ginover, Guenore, Wenneveria and Winlogee, among other variants. Most of the time, I have called her Guinevere. My system, which is rough and ready, breaks down with Tristan or Tristram. He looks absurd as Tristram in French and German contexts, and not very happy as Tristan in English ones. I have therefore used both forms of the name.

My debt to other people's books, listed in the Bibliography, is immense. In many cases, however, I have approached the material from a different point of view. I do not believe that the peculiar force of the Arthurian stories stems entirely from their Celtic background and owes practically nothing to medieval writers. Nor do I accept the contrary proposition, that the Celtic background is of little significance. Arguments about origins and influences, though important and fascinating, sometimes obscure the stories themselves.

I

Arthur of Britain

> When Arthur reached a field of battle bright
> With pitched pavilions of his foe, the world
> Was all so clear about him that he saw
> The smallest rock far on the faintest hill,
> And even in high day the morning star.
>
> Tennyson, *The Coming of Arthur*

Arthur has been a name to conjure with for fifteen centuries. Legend places him in a remote and misty past as an illustrious British king or emperor, who conquered most of western Europe. He ruled his dominions from his capital city, many-towered Camelot, where he held his glittering court. His adviser was Merlin, the master magician. He founded the Order of the Round Table, the bravest fellowship of chivalrous knights ever to dominate a battlefield and support a throne. In a blazing splendour of helm and banner. lance and shield, the knights rode out to seek adventure and glory, notably in the quest of the Grail, the holiest relic in Christendom.

Arthur's brilliant career was doomed to a tragic end. His beautiful wife, Guinevere, and his dearest friend and champion, Lancelot, fell passionately in love. Their relationship ruined Arthur's happiness, plunged his kingdom into civil war and destroyed the Round Table. While the great king was struggling to salvage what he could from the wreck, he was betrayed by his bastard son, Mordred. Fatally wounded by Mordred in his last battle, Arthur was carried away to Avalon, the paradise in the west from which he will return to lead his countrymen at the time when they most have need of him.

Arthur and Merlin, Guinevere and Lancelot, Gawain and Galahad,

Tristram and Yseult cast a golden romantic spell to this day. In the middle ages, however, when the stories of the Round Table took shape, Arthur and his court were thought of as real people. They had lived and loved and died at some uncertain period in the past, but in medieval conditions. They had lived close to valour, beauty, love and luxury, but also cheek by jowl with pain, fear, sorrow and death. The tales of their adventures were exciting and enjoyable, simply as adventures, but since they were about real people they were also instructive and inspiring. They showed how men and women could best live their lives in an imperfect world. These are qualities which the Arthurian legends have never lost.

Arthur and his knights burst on to the European literary scene in the twelfth century and the tales of the Round Table swiftly became the favourite entertainment of the leisured classes in France and England, Germany, Italy and Spain. One French writer said that Arthur's name and fame were known as far as the empire of Christendom extended, from Britain to the Middle East. Another thought there were only three subjects worthy of an epic poet's attention: the Matter of France, the Matter of Britain and the Matter of Rome. The Matter of France meant the exploits of the Emperor Charlemagne and his paladins, and the Matter of Rome consisted principally of the legends of Troy, Aeneas and Alexander the Great. They were supplanted in popular favour by the Matter of Britain, the stories of Arthur and his knights.

The Matter of Britain is largely the creation of medieval writers and medieval French culture, but its history goes much further back into Celtic legend and mythology. At the root it belongs to the same class of 'heroic' literature as the *Iliad*, the *Nibelungenlied* and the Norse and Irish sagas. A hero, in this sense, is first and foremost someone pre-eminent in courage, strength and skill in war. Usually he comes from the ruling class, who specialize in war. Heroic literature grows up in societies where people live in comparatively small groups, dominated by a warrior aristocracy and engaged in constant raiding and warfare. The warfare is on a small scale by modern standards. It is the type of fighting in which individual bravery and skill are bound to be noticed and are likely to have a decisive effect on the result. The two qualities which a society of this kind most needs in its fighting men are courage and loyalty, and these are the essential qualities of a hero.

Since the hero's staple occupation is fighting, his life is likely to be short. He knows this, and what spurs him on to live well and die well is the longing for fame, to be honoured in song and story for generations after his death. Achilles, the greatest hero of the *Iliad*, deliberately chose a brief life of valour and undying renown to a long, quiet life in obscurity. The same choice was made in Irish legend by Cuchulain, the supreme hero of Ulster. As a boy he heard a Druid predict that a young warrior who was formally presented with his weapons on that day would win immortal fame, but would die young. Cuchulain thought it a fair bargain. He demanded and was given his arms on the spot.

Though he is a formidable fighter, the hero must not be a brute. He needs the gentler qualities and a nobler purpose than to kill for the love of killing, or to acquire power and wealth. He is loyal to a leader, to people for whom he is responsible, to an ideal. A king's bodyguard defends him to the death. David risks his life against the giant Goliath to protect the Israelites. Beowulf is killed saving his people from a monstrous dragon.

The hero's guiding star is his ideal of what he ought to be. Hector leaves the safety of Troy and stalks out to fight Achilles, knowing that his death is certain, but preferring to be killed than to be a coward. The blinded Samson, chained and mocked, pulls down the pillars of the hall to take his foes with him into the night of death. Ragnar Lothbrok, the Viking chieftain, cast by his enemies into a pit of snakes, dies with a defiant song on his lips, refusing to ask for mercy.

The true hero rises above the ordinary human levels of courage and resolution, strength of character and physical strength, generosity and greatness of soul. He seems superhuman: so much so that he may be regarded as semi-divine. The Greeks prayed for help to famous dead heroes at their shrines, where their relics – their bones or anything closely associated with them – were carefully preserved and honoured. Catholic Christians, similarly, have for centuries venerated the relics of saints and prayed to them for help.

Legends of heroes do not all tell fundamentally the same story, and attempts to show that they do end in twisting them out of shape. However, some patterns do occur fairly frequently. In one of them, which often appears in Arthurian tales, the hero leaves the world of men and ventures into an otherworld, a strange and perilous region, which is the realm of gods, spirits and uncanny beings. In the otherworld he performs some great exploit or acquires something of inestimable value. He may kill a dragon or a monster; he may set prisoners free and restore them to life in the human world; he may acquire wisdom or a beautiful bride or a priceless treasure. He may even, as in the Grail legends, discover the ultimate secret of life and win immortality.

The Babylonian hero Gilgamesh went to rescue his dearest friend from the land of death, and he had the plant of immortality within his grasp when a serpent stole it from him, preventing him from bringing back to men the gift of eternal life. Hercules killed the dragon which watched over the golden apples of immortality in the Garden of the Hesperides, and carried them off, though they had afterwards to be restored to the garden by the gods. Theseus went to Crete and penetrated the mysterious labyrinth, the lair of the Minotaur. By killing the monster he freed his fellow prisoners from a sentence of death, and ended the evil custom of the annual tribute of young men and girls sent from Athens to Crete to be devoured by the Minotaur. Jason and the Argonauts sailed to the far-off uncanny country of Aeëtes, the wizard-king of Colchis, where the golden fleece was guarded by a dragon. Jason killed the dragon and carried off the fleece.

In some stories the hero stays in the otherworld for ever, lost to the world of men, but if he is to be in the front rank of heroes he must return to the human world. He must bring back the spoils of his victory, the fruits of his experience, for the blessing and benefit of his fellow men: whether he is Moses coming down from Mount Sinai with the tablets of the law or Prometheus stealing fire from heaven and bestowing it on mankind. The hero brings back something that the world lacks and yearns for, something which regenerates and enhances life.

It seems likely that this pattern goes back ultimately to the practices of the shamans, or priest-magicians, of prehistoric tribes. The shaman put himself into trances, in which he visited the world of gods and spirits, where he encountered terrifying dangers and obstacles. On these expeditions he rescued the souls of the dying from the otherworld and restored them to life in their human bodies. He also gained knowledge of the mysterious forces which rule nature and govern life and death, and he used his knowledge for the benefit of his tribe.

The shaman's trance-experiences support the modern psychological interpretation of heroic legends, in which the otherworld that the hero invades is his own unconscious mind. It is there, in the darkest and most perilous regions of human nature, where the springs of character and action lie, that he wins his victory and discovers truth. What he discovers is himself, and he returns to the everyday world with 'the treasure hard to attain'. The treasure is integrity.

The hero is Everyman carried to a higher power, and the life-enhancing gift which he brings to men is his inspiring example. He confers dignity on the human race by showing that life need not be petty. The limits of human reach and achievement are not as narrow and restricted as they may seem. Existence can be vivid and exciting. Whether in history or in fiction – and the gulf between the two is not as wide as is often supposed – the hero upholds the possibility of living a life which is richer, more intense and more complete than is the common lot of man.

The Real Arthur

Although the legends of Arthur are fictitious, it is difficult to account for their existence unless there was originally a real man whose character and achievements formed the rock on which the towers and battlements of legend were raised. There is no certainty that a real Arthur existed, but he probably did. Little evidence has survived from his time, but we do know that in the fifth century the Roman Empire in the West collapsed and the Roman province of Britain was threatened with conquest by the Saxons. After years of war, the Roman-British defeated the Saxons, in about AD 500, in a battle decisive enough to restore peace and stifle the Saxon menace for half a century. It is not unduly far-fetched to suppose that the British army which won the decisive battle had a leader. If so, he was the original of the legendary King Arthur.

This means, ironically, that the great British hero fought against the ancestors of most modern Britons. He was probably not a king but a British aristocrat with a heritage partly Roman and partly Celtic. His name was Artorius, which is a Roman surname (perhaps a Latinized form of a Celtic name meaning 'bear'). It is possible that he was a leader of cavalry and that his victorious horsemen became the mounted knights of the medieval legends. It is also possible, though this is highly speculative, that after his defeat of the Saxons he imposed his authority on the whole of England and Wales.

Britain was part of the Roman Empire for more than 350 years, roughly the same length of time as separates us now from the Pilgrim Fathers, King James I and Cardinal Richelieu. It was a period long enough to make Roman rule in Britain seem part of the natural order of things and its collapse in the face of barbarian invasions all the more shocking. This in turn gave Arthur's achievement in subduing the barbarians and restoring order a glow of heroic magnificence which was the basis of his legendary fame.

The Romans governed England and Wales with their northern frontier, most of the time, on Hadrian's Wall. Roman Britain was peaceful and prosperous. The leading native Celtic families adopted Roman customs, manners and attitudes, and in time also adopted Christianity. The new faith may have reached the island in the first century, brought in by soldiers and traders. In the fourth century Christianity became the official religion of the Empire. By then it was well established in Britain, but the older pagan cults were still flourishing side by side with it, and as late as the 360s a fine new temple was built for the Celtic god Nodens at Lydney in Gloucestershire. A good many Britons, probably, were not exactly either Christians or pagans, but a mixture of the two.

From late in the third century the wealthy Roman province became a magnet for savage raids by the Irish from the west, the Picts from Scotland in the north, and the Saxons from the east. Most of the attackers were bandits, not immigrants. They came and saw and plundered, and sailed home again with their loot. But many Irish families settled in Wales and some Saxons may have settled down in eastern England. The raids continued into the fifth century.

Meanwhile Roman power on the Continent came under irresistible pressure from barbarian invaders. The Goths crossed the Danube and in 378 annihilated a Roman army and killed the Emperor Valens. In 406 an enormous horde of Germans swarmed across the Rhine and descended on France. They could not be driven out but were with difficulty ejected into Spain and North Africa. The city of Rome itself was sacked by the Goths in 410. The Empire in the West fell to pieces. The British were left to fend for themselves.

The course of events is obscure, but it seems that a ruler called Vortigern, who originally came from Wales, established himself in command of most of Roman Britain. Tradition blamed him for the disaster which followed. For defence against the Picts he hired a war-band of Saxons, led by a chieftain named Hengist, and encouraged other Saxons to settle in eastern England. Presently, like cuckoos in the nest, the Saxons turned against their hosts. It is simpler to call them Saxons, for clarity, but they called themselves the English.

The horror with which civilized Roman-Britons regarded these primitive savages was expressed a century later by the British writer Gildas, who calls them 'the vile unspeakable Saxons, hated of God and man alike'. For a Christian like Gildas the horror was deepened by the fact that the Saxons were pagans. He looked back to the Saxon revolt with quivering revulsion, especially to the profanation of Christian values and the destruction of town life, one of the hallmarks of Roman civilization.

> All the greater towns fell to the enemy's battering rams; all their inhabitants, bishops, priests and people were mown down together while swords flashed and flames crackled. Horrible it was to see the foundation stones of towers and high walls thrown down bottom upward in the squares, mixing with holy altars and fragments of human bodies, as though they were covered with a purple crest of clotted blood, as in some fantastic wine-press. There was no burial save in the ruins of the houses, or in the bellies of the beasts and birds.[1]

Many Britons fled from the Saxon terror to the Continent and settled in Normandy and Brittany. Others resisted, led by a nobleman named Ambrosius Aurelianus and subsequently, it seems, by Arthur. The struggle swung this way and that for years until eventually the British, probably under Arthur, crushed the Saxons with great slaughter in a battle at Badon Hill. The exact locality of Badon Hill is uncertain, but it was somewhere in southern England. The Welsh Annals (compiled about 950) give the date of the battle as 516, but this is probably about sixteen years too late. The victory ended the Saxon threat for fifty years or more. The savages stayed quietly on their reservations in the east and south-east, until they erupted again two or three generations later.

The only early account of these events is in Gildas's book, written about 540. He says that the Saxons were decisively defeated at Badon, but unfortunately he does not mention Arthur. This has naturally suggested to sceptical historians that Arthur never existed. On the other hand, Gildas was not writing a history of the past but a polemic against his own contemporaries, and he generally avoided mentioning personal names in any case. He grew up in the period of peace after the victory at Badon, which he says was a gift of God that made possible the preservation of civilized order and good government. He does not say who won the victory, but somebody did, and the obvious candidate is Arthur.

In the century after 500 several British kings and princes are known to have named their sons Arthur, apparently in salute to the hero. In about 600 the famous Welsh bard Aneirin wrote a poem, the *Gododdin*, in which he says of one of his characters: 'he glutted black ravens on the rampart of the city, though he was not Arthur'.[2] This reference to Arthur as a redoubtable warrior, slaughtering many opponents, ought to leave no doubt that he existed, but the phrase 'though he was not Arthur' might be a later interpolation.

There is then a long gap of two hundred years to a history of the Britons compiled in about 800 by a Welsh monk, Nennius. He says that the Saxon chief Hengist was succeeded in Kent by his son Octha. 'Of him sprang the kings of the Kentishmen. Then Arthur fought against them in those days with the kings of the Britons, and it was he who led their battles.' Arthur is not described as a king here, but as *dux bellorum*: war-leader or commander-in-chief.[3]

Nennius gives a list, apparently drawn from a Welsh poem, of twelve victories which Arthur won, many of them at places which cannot now be identified. Four of the battles occurred in the region of Linnuis, which is presumably modern Lindsey, the area round Lincoln. Another was fought in 'the wood of Celidon', the Caledonian Forest somewhere in the lowlands of Scotland. The ninth battle was at the City of the Legion, which must be either Chester or Caerleon. The twelfth was the victory of Badon Hill, where Arthur, apparently single-handed, killed almost a thousand of the enemy in one charge. Legend had already made him larger than life.

Nennius comes too late to prove that Arthur really lived, though he does show that Arthur was now believed to have been the British war-leader who overwhelmed the Saxons at Badon. The other battles may or may not have been fought by the real Arthur. If they were, he covered a great deal of ground in his military career and he must have engaged other opponents besides Saxons in the south and the midlands: possibly Picts, Irish and recalcitrant Britons.

It is clear from Nennius that folklore had now grown up about Arthur in Wales. A stone marked with the print of a dog's paw could be seen on top of a cairn in Breconshire. It was said that the dog was Arthur's hound, Cabal, which trod on the stone during the hunting of the boar named Troit (of which more later). Arthur made the cairn and if a stone was taken away from the pile, it always returned. Another story was that Arthur killed his own son, Amr, and buried him in Herefordshire. The grave mound mysteriously varied in length, from nine to fifteen feet. Local traditions of this kind, linking the hero with odd or striking features of the countryside, later multiplied enormously. Arthur is now connected with more hills, rocks and natural features in Britain than anyone else except the Devil.

The war-leader may have had a principal base and headquarters, which in the legends became the beautiful medieval city of Camelot, Arthur's capital. Camelot is first mentioned in the 1170s, by Chrétien de Troyes in his *Lancelot*. The name may have been adapted from Camulodunum, the Roman name of Colchester. Alternatively, in Cornwall the river Camel and the town of Camelford are not far from Tintagel, Arthur's legendary birthplace. In the fifteenth century Malory identified Camelot as Winchester in the *Morte Darthur*, but Caxton, who printed it, seems to have thought that Camelot was either Caerleon or Caerwent. In the sixteenth century the antiquaries Leland and Camden located Camelot at Cadbury Castle in Somerset. Cadbury is twelve miles from Glastonbury, which has the strongest associations with Arthur and the Grail of any place in Britain. The 'castle' is not a medieval fortress but an earthwork fort, originally pre-Roman, on a hill near the river Cam and the village of Queen Camel, names which recall Camelot. Local tradition going back to the sixteenth century and beyond called the fort Arthur's Palace. Excavations at Cadbury since 1966 have shown that in about 500 the hill was the stronghold of a powerful chieftain, who commanded sufficient resources of men and money to improve the defences on a scale so far unknown elsewhere in Britain at this period. There is no definite evidence that this chieftain was the real Arthur, but it is an enticing possibility.

Arthur's death is recorded in the Welsh Annals, which report the battle of Camlann, in which 'Arthur and Medraut fell'. The date is given as twenty-one years after the battle of Badon Hill, which would put it at about 520. The Annals do not say that Arthur and Medraut were on opposite sides, but Medraut is the treacherous Mordred of the later legends. Camlann might be Camboglanna, a fort at Birdoswald on Hadrian's Wall, but again the rivers Camel in Cornwall and Cam in Somerset are possible sites. Cornish tradition places the battle at Slaughter Bridge, on the Camel near Camelford. On the other hand, the Somerset Cam is close to Glastonbury, which the legends connect with Arthur's last battle and his passing to Avalon.

The real Arthur, then, was the general who won the crushing victory at Badon which turned back the Saxon tide. He may have had his headquarters at Cadbury. He probably fought other battles, including the one in which he was killed in about 520. William of Malmesbury, a tough-minded Norman historian of the twelfth century, wrote: 'It is of this Arthur that the Britons [or Bretons] fondly tell so many fables, even to the present day; a man worthy to be celebrated, not by idle fictions, but by authentic history.'[4]

Nennius describes Arthur as the war-leader of 'the kings of the Britons'. In other words, he commanded an army made up of leading British chieftains and their troops: hence the picture of him in legend as a great king and warrior, admired by lesser kings, princes and fighting men who flocked to serve under him. His victory was followed by a period of peace, order and British supremacy over the Saxons which later generations of Britons looked back on as a golden age.

If Arthur headed or supported an effective central government, it had broken down twenty years or so after his death, when Gildas was writing. Britain was then a patchwork of quarrelling kingdoms, ruled by local despots. Gildas denounced them for bringing anarchy back to Britain and for living by terror and plunder. He predicted that the vile and unspeakable Saxons would rise again, and time proved him right. There was a new wave of Saxon expansion in the second half of the century. By about 600, Saxon leaders had overrun most of England.

The Britons retained their independence only in Wales, in Cornwall and the south-west for a time, and north of Hadrian's Wall. The Saxons called them 'foreigners', *wealh*, which in modern English is 'Welsh'. They called themselves 'fellow-countrymen', *combrogi*, whose modern Welsh form is *Cymry*, and from which Cumberland and Cumbria are also derived. They never forgot Arthur and his lost age of British supremacy. The tales they told of him were not really 'idle fictions', for he became a symbol and focal point of their national identity and pride.

Arthur in Celtic Legend

The name Celt may come from a word meaning 'to fight'. The earliest people who can be identified as Celts lived in central Europe, around the upper reaches of the Danube and the Rhine. Over many centuries groups of them moved north and west into northern Germany, France, Spain, Britain and Ireland. Others invaded Italy and raided Rome in 390 BC. Still others went south-east into Greece, where they plundered Delphi in 278 BC, and eventually settled in Asia Minor. In early times they were a horse-taming people, equipped with wagons and war-chariots, and they were also head-hunters. Greek and Roman writers described them as great lovers of fighting and drinking, high-spirited, boastful, ostentatious, reckless and impetuously brave.

In Celtic society the farmers who raised cattle and grew crops were protected by an aristocracy of kings, lesser chieftains, warriors and priests. The aristocrats were the patrons of skilled craftsmen – armourers, goldsmiths, jewellers – and of artists with words, the bards. The principal occupation of the aristocracy was war, and the principal duty of a bard was to compose and recite stirring poems about the exploits of warriors and the distinguished ancestry and achievements of the king or chieftain who was his patron.

When the Romans conquered Britain they imposed on Celtic tribal kingdoms organized for war the tranquillizing influence of the Roman peace. When Roman rule ended and the struggles against the Saxons began, warfare once more became part of normal life and the native Celtic pattern of society re-emerged in the areas where the Celts retained their independence. It had persisted all along in Ireland, which the Romans never reached (and something close to it survived in the Scottish Highlands down into the eighteenth century). This social pattern accounts for many of the characteristics of the Matter of Britain.

Each king or chieftain had his war-band, formed by his own male relatives and other fighting men of good birth and breeding who took service with him. He supplied them with food, drink, lodging and entertainment in his hall, with weapons and loot, and sometimes with land. In return he demanded their loyal support in war and in peace. The warrior's compensation for a life which was likely to be short was to be held in honour in his own lifetime and ideally for generations after his death. He hoped to earn renown by his exploits in battle and in his other occupations of raiding, cattle-rustling, hunting and drinking. 'Though they were slain they slew,' says the *Gododdin* of its heroes, 'and they shall be honoured till the end of the world.'[5]

The *Gododdin* is a lament for the retinue of Mynyddog the Wealthy, who in about AD 600 ruled south-eastern Scotland and north-eastern England from his capital at Edinburgh. Recruiting fighting men from Wales and from the Picts in the north to strengthen his war-band, he launched a force of three hundred cavalry against the Saxons to the south. The Saxons were still fiercely detested as savages, pagans and 'mongrels'. The British horsemen met the enemy at Catraeth (Catterick in Yorkshire). They knew they had little hope, but they were true to their salt and fought against impossible odds to the death. Only one man escaped to tell the tale.

The warrior's second compensation for a short life was a merry one. He was free of the ignoble necessity to work. Successful chieftains provided plentiful food and drink in their halls, ravishing poetry, fine weapons and large quantities of looted treasure. The successful leader needed to be active, brave and skilful in war, and hospitable and open-handed at home.

Arthur is the ideal leader of this kind in the Welsh tradition, the bravest and best of fighters and commanders, the most generous of hosts. The *Gododdin* uses his name as a byword for warlike prowess. Another Welsh poem, on the death in battle of a sixth-century king, Geraint son of Erbin, mentions 'bold men of Arthur's who hewed with steel' and calls him 'the emperor, the ruler in the toil of battle'. In the story 'Culhwch and Olwen', the hero comes to Arthur's hall and, although he is too late to be allowed in that evening, the gatekeeper tells him to go to the guest-house, where he will be well looked after.

> Meat for thy dogs and corn for thy horse, and hot peppered chops for thyself, and wine brimming over, and delectable songs before thee. Food for fifty men shall come to thee in the hospice; there men from afar take their meat. . . . It will be no worse for thee there than for Arthur in the court: a woman to sleep with thee, and delectable songs before thee.[6]

For its survival a heroic society of this kind requires in its fighting men a combination of individual prowess and pride with loyalty to a leader and an ability to live in a rough and ready harmony with the rest of the war-band. The Celtic tales which have survived focus mainly on the fierce individualism of the warrior, his dauntless courage, iron resolution and adamant pride, and they obviously exaggerate this side of the equation at the expense of the other.

The heroes of these stories find a fulfilling joy in battle. Like the knights in medieval Arthurian tales, they are only fully themselves in a fight. Only then are their qualities of courage, skill and strength employed to the utmost, their weaknesses forgotten, their misgivings and self-doubt stilled. This is why they love fighting. In it they find release from the prison of their own inadequacies and they attain a rapturous animal ferocity. 'The hero red in his fury,' says the *Gododdin*, 'the man-slaying champion, was wont to be joyful like a wolf at his post, the wolf of the army.'[7]

At the same time, the *Gododdin* is shot through with a sense of the tragedy of war. 'After wine-feast and mead-feast they went from us, the mail-clad warriors. I know the grief for their death. Their slaying came to pass before they could grow grey-haired, their host was high-spirited in front of Catraeth.' Or again: 'In battle they made women widows, and many a mother with tears at her eyelids.'[8] There is a feeling of inevitability, of a circular process at work. It is tragic that the flower of heroism should be cut down, but without the prospect of early death in battle, the flower would never have grown.

With the love of battle went a sense of fairness. When a champion offered single combat, it was considered wrong for more than one of his opponents to attack him. This rule was known in Ireland as *fir fer*, literally 'fair play'. In the Irish epic *Tain Bo Cuailnge* (The Cattle Raid of Cooley), the hero Cuchulain holds up an enemy army and gains time for the Ulstermen by fighting single combat after single combat at a ford. The enemy could overwhelm him by attacking him in force, but they do not. Eventually they send against him his dearly loved friend Ferdia mac Damain. The two men grew up together as boys and they are profoundly reluctant to fight, but honour compels them. Cuchulain says sardonically to Ferdia: 'My bosom friend and heart's blood, dear above all, I am going to miss you.' Ferdia answers: 'You make much of yourself, but the fight is to come. I'll have spiked your head when the cock crows.' The combat is fought out with heart-breaking courtesy until Ferdia is killed and Cuchulain is plunged in grief.[9]

The ideal warrior was expected to have other talents beside a gift for fighting. Cuchulain possessed 'many and various gifts: the gift of beauty, the gift of form, the gift of build, the gift of swimming, the gift of horsemanship, the gift of playing *fidchell*, the gift of playing *brandub*, the gift of battle, the gift of fighting, the gift of conflict, the gift of sight, the gift of speech, the gift of counsel, the gift of fowling, the gift of laying waste, the gift of plundering in a strange border'.[10] *Fidchell* and *brandub*, 'wooden wisdom' and 'black raven', were board-games.

The hero who was as fierce as a wolf in battle was expected to be courteous and pleasant at home, and gentle with women. In real life, very often, these ideals were not lived up to: ideals never are. Rules of fair play were not always observed. Mead-feasts and wine-feasts did not invariably make for pleasant comradeship. In drunkenness the warrior could find something of the same liberation from himself, something of the same rapture, which he discovered in battle, and there are plenty of stories of the quarrelsomeness and touchy pride of the war-band in its own hall. The ideals are there, all the same, and they are closely related to the ideals of knightly chivalry which underlie the medieval Matter of Britain.

Much of the land-mass of Welsh Arthurian legend has sunk below the surface. Islands and isolated rocks are still visible here and there, and more of the lost territory can sometimes be dimly made out by looking down through the waters of time which have covered it. In one poem, in the Black Book of Carmarthen, Arthur arrives at his hall and his gatekeeper asks who is with him. Arthur replies with a list of his war-band. Among them are Cai (or Cei) and Bedwyr, who later became two of the leading knights of the Round Table, Sir Kay and Sir Bedivere.

An army was vanity
Compared with Cei in battle...
When he drank from a horn
He would drink as much as four;
Into battle when he came
He slew as would a hundred.
Unless God should accomplish it,
Cei's death would be unattainable.[11]

There are tantalizingly cryptic references to Arthur and his men struggling with a hag in the hall of Afarnach and fighting dog-headed beings on 'the mountain of Eidyn', which is the hill at Edinburgh now known as Arthur's Seat. Cai is said to have pierced nine witches, destroyed lions in Anglesey and polished his shield to deal with Palug's Cat (a monster which later turned into the Demon Cat of Lausanne). Cai's polishing his shield suggests a parallel with the Greek legend of Perseus and the Gorgon, the sight of which turned all things to stone. Perseus cut off the sleeping monster's head in safety by looking not at her but at her reflection in his polished shield.

Arthur, the destroyer of Saxons, has here moved on to a semi-supernatural plane as a hero who rids the land of evil creatures, witches and monsters. Besides Cai and Bedwyr, his war-band includes men who were originally Celtic gods: Mabon son of Modron, servant of Uther Pendragon; Manawydan son of Llyr, 'profound in counsel'; and Llwch Llauynnaug. The name Mabon son of Modron means 'Son son of Mother'. He was identified with Apollo by the Romans, who called him Apollo Maponos: Apollo the Young God. Manawydan son of Llyr was a sea god, known in Ireland as Manannan mac Lir. The Isle of Man is named after him. Llwch was the god known to the Irish as Lugh. He was the divine father of Cuchulain. The close connection between Welsh and Irish mythology is natural, not only because both peoples were Celts but because there was constant communication between them across the Irish Sea.

By the time of this poem the Welsh were Christians, but the pagan gods lingered on as superhuman beings who could suitably be located in the same vaguely remembered and glorious past as Arthur, and could be made into his subordinates. As they had moved down the ladder from god to semi-divine hero, so Arthur had moved up it and his attraction was so strong that, like a magnet, he drew into his field stories and characters not originally connected with him.

Arthur again moves on a supernatural plane in 'The Spoils of Annwn', a poem in the Book of Taliesin. The date and authorship of the poem are unknown. It is about a raid on the otherworld, the island fortress or city of Annwn. Arthur sailed there in his ship Prydwen (Fair Face) and, although he took three times the ship's normal complement with him, only seven men returned from the expedition alive.

The object of the raid was to seize the magic cauldron of Annwn, from which only the brave could eat, for it would not boil food for a coward. It had pearls round its rim and was kindled by the breath of nine maidens. In the poem Annwn is also called Caer Wydr, the Fortress of Glass; Caer Feddiwid, the Fortress of Carousal, because lavish feasting was one of the pleasures of the Celtic otherworld; and Caer Siddi, the Fairy Fortress. The *side* are the prehistoric grave mounds in Ireland, which traditionally are entrances to the uncanny world of gods, spirits and the fairy people.

The story belongs to the genre in which the hero invades the otherworld to carry off a life-enhancing treasure. Celtic otherworlds frequently contained magic vessels of plenty and regeneration. The cauldron of Annwn is probably one of them, and is therefore close to the origins of the Grail. Whether Arthur succeeded in seizing it is not clear, but presumably he did. Annwn is described only allusively, but it seems to be both an alluring and a dangerous place. A mysterious prisoner named Gwair is chained there. Six thousand warriors line the walls of the Glass Fortress and their sentinel will not speak when he is hailed. Time stands still there, or is different from mortal time, for it is pitch dark at noon. The poem mentions a sword, apparently brought back from Annwn, so there may already have been a story that Arthur obtained a magic sword from the otherworld.

Celtic traditions of the otherworld had a powerful influence on the medieval Arthurian stories. The otherworld is an enchanted place, beautiful and uncanny. It does not obey the rules of the human world. It is one place and also many places, far away and close at hand. It may be an island or a group of islands out to sea in the west. It may be under the ground, at the bottom of a lake or inside a hill or a burial mound. It may be just round the next corner, though you will not usually see it there, for it is normally veiled from mortal eyes.

In the otherworld there is no death and no time. No one grows older there, no one falls ill and no one works. Food and drink are supplied by magic cauldrons and vessels of various kinds. The people of the otherworld spend their time in delightful idleness, feasting and lovemaking. They have beautiful music to hear and beautiful scents to smell. The flowers bloom all the year round and the trees are always richly hung with fruit.

The entrancing otherworld is dangerous for mortals, precisely because it is so attractive. Psychologically, it is a territory in the mind. It is the world of dream and fantasy, where the normal rules do not apply and which in its own way is one place and many places, near at hand and far away. A hero who enters it may not find it easy to escape again. He may be caught in a web of sensuous pleasure and slothful ease, in which his energies are sapped and his powers of decision melt away. The parallel in real life is the danger of living in a world of daydream and fantasy which makes it impossible to cope with everyday reality. The theme appears frequently in medieval Arthurian stories of knights taken prisoner in enchanted islands or castles.

The idyllic islands of Celtic mythology in many ways resemble Greek descriptions of the Fortunate Isles or Isles of the Blessed, the home of the dead over the sea to the west. In many Celtic stories, however, it is not clear whether the otherworld is the land of the dead or not. It is sometimes the home of the gods and very often it is inhabited by a fairy race, happy and beautiful and for ever young.

The theme of a raid on the otherworld recurs in disguised form in 'Culhwch and Olwen' (in *The Mabinogion*), a story bubbling over with charm and exuberance. The story probably reached its present form by 1100. How much further it goes back is not known, but the episode of Arthur hunting the great boar Twrch Trwyth was in circulation by about 800, because Nennius mentions it in passing as the hunting of the boar Troit. The central thread of the story is the hero's quest for the giant's daughter, which is a widespread folk-tale theme. Round this are woven numerous motifs and characters from mythology and folk belief, including some from Irish traditions.

The hero, Culhwch, is Arthur's first cousin. He rides to Arthur's court to seek help in the task, imposed on him by his wicked stepmother, of winning the hand of the beautiful Olwen, daughter of Ysbadadden, the chief giant. Ysbadadden knows that he will die when his daughter takes a man. He is therefore determined to keep her a virgin, and none of her suitors has ever left the giant's fortress alive.

When Culhwch reaches Arthur's court, the gatekeeper refuses to let him see Arthur at once, because it is too late in the evening. Culhwch threatens to make all the pregnant women at the court miscarry by emitting three frightful screams at the gate. The gatekeeper goes in to consult Arthur, and Cai, here showing the officious and curmudgeonly character which is typical of him in the later stories, objects to letting Culhwch in against the rules of the court. But Arthur, who is shrewder and more generous, tells Cai that a man is only noble so long as people make demands on him.

Arthur is now definitely King Arthur, 'the sovereign prince of this island', and he dominates Ireland, Brittany, Normandy and the rest of France as well as Britain. The gatekeeper at one point makes a boastful speech in which he mentions Arthur's conquest of Greece. The picture of Arthur as the conqueror of much of Europe is already emerging, and the reference to Greece may perhaps be a distant reminiscence of the Celtic drive through Greece in the third century.

Culhwch is ushered in and asks for Arthur's help, which is readily promised. He then recites an immensely long list of Arthur's war-band, well over two hundred of them. This is a magical incantation in which the names of Arthur's warriors are used as words of power. Culhwch invokes his intended bride, Olwen, in the names of Arthur's men, thereby magically bringing their energy and prowess to bear on his quest for her.

Cai and Bedwyr are named first of the war-band, and more information is given about them later in the story. Bedwyr was one of the three handsomest men in Britain (Arthur himself was one of the others) and he had only one hand, though no warrior drew blood in battle faster than he did. Cai had various peculiarities. He could go without sleep for nine days and nights, and could hold his breath under water for the same length of time. No doctor could heal a wound made by his sword. When he wished he could make himself as tall as a tree. His heart and his hands were cold and yet he could give out such heat that his comrades lit fires with him: this apparently contradictory peculiarity seems to follow from his link with trees.

Manawydan and Llwch reappear in the war-band. With them now is another former deity, Gwynn son of Nudd, in whom God locked up the power of the demons of Annwn to prevent them from destroying the world. In Welsh folklore Gwynn was the lord of the otherworld and the king of the fairy people. He had a palace on top of Glastonbury Tor.

Also in the war-band are Arthur's nephew Gwalchmei (later Sir Gawain), Drwst Iron Fist (probably Tristram) and Llenlleawg the Irishman (Lancelot). Gildas, the sixth-century author, is included and so is Echel Pierced Thigh, who seems to have been Achilles originally. Several names on the list come from Irish legends. Some of the war-band are personifications of superhuman qualities. Drem son of Dremidydd (Sight son of Seer) has such preternatural sharpness of vision that he can see a gnat in Scotland from as far away as Cornwall.

Clust son of Clustfeinad (Ear son of Hearer) can hear an ant stirring fifty miles off.

Arthur's wife is Gwenhwyfar (the Welsh form of Guinevere, possibly meaning 'white phantom' or, in effect, 'white goddess'). The famous poet Taliesin is his chief bard and his court magician is Menw son of Teirgwaedd, a skilful caster of spells who can turn himself into a bird, which suggests that he is really a Druid. Arthur's sword is Caledfwlch (later Excalibur). His ship is Prydwen and his great hunting dog is Cafall (Horse), who is also mentioned by Nennius. His headquarters are in Cornwall at Celliwig, which may be Kelly Rounds, a hill fort near Padstow, at the northern end of the old trade route across Cornwall from Brittany to Wales and Ireland.

Arthur is now at the head of a band of redoubtable warriors, drawn from a wide range of traditions. Some of them reappear later as knights of the Round Table and, as in the later tales of the Round Table, Arthur's court is the centre from which the heroes go out to seek adventure. But Arthur and his men are still comparatively primitive figures, products of a simpler society than the world of feudal chivalry to which they were afterwards adapted. There are few traces of Christianity in 'Culhwch and Olwen' and its atmosphere is almost entirely pagan. There are no bishops or priests at Arthur's court, but there does seem to be a Druid. Generous, brave, hot-blooded, boisterous and aggressive, the king and his men rampage about the countryside putting down supernatural foes – giants, witches and monsters. They themselves have supernatural powers and they are as unpredictable and uncontrollable as pagan gods. They are regarded as benevolent heroes who protect society from evil and destructive forces, which was later the role of the knights of the Round Table.

Arthur sends some of his men with Culhwch to find the giant Ysbadadden and after a long search they come to his stronghold. The giant is so huge that his servants have to push his eyelids up with forks when he wants to see his visitors, and so primitive that he is armed with stone spears. After several unsuccessful attempts to kill his unwelcome guests, the giant imposes on Culhwch forty impossible tasks which he must carry out to win Olwen.

These tasks are *anoetheu*: wonders, marvels, things hard to achieve or obtain. Among them are the comb and scissors from the head of the boar Twrch Trwyth, with which Ysbadadden will dress his hair for Olwen's wedding, and the blood of the Black Witch, which the giant will use as shaving-soap. Culhwch must also find and bring back four magical vessels to supply food and drink for the wedding feast. The first is the cup of Llwyr, which provides the best and strongest of all drink. The second is the plate or table of Gwyddneu Long Leg, which supplies limitless quantities of food, and to each person the food he likes best. The giant intends to eat from it on the night when Culhwch sleeps with Olwen, presumably to stave off the death which will otherwise overtake him. The third is the drinking-horn of Gwlgawd, from which to pour the drink, and the fourth is the cauldron of Diwrnach the Irishman, in which to boil the meat for the feast.

These are almost certainly otherworld vessels. Supplying delectable food and drink which staved off death was later a characteristic of the Grail. The cauldron of Diwrnach may be the cauldron of 'The Spoils of Annwn', for a much later list of the Thirteen Treasures of the Island of Britain includes the cauldron of Tyrnoc, which like the cauldron of Annwn would not cook food for a coward. There seems to be a hint at the cauldron's magic regenerative powers in 'Culhwch and Olwen', when Diwrnach the Irishman says that Arthur shall not have the cauldron, though he might be the better for a mere glimpse of it.

Arthur and his men sail to Ireland in Prydwen. They carry off the cauldron and bring it back, stuffed with treasures, to Dyfed (south-west Wales). This is apparently another version of the tale of Arthur's otherworld raid in 'The Spoils of Annwn', with the otherworld here located in Ireland.

Arthur and his warriors also seek out the Black Witch in her cave and Arthur, told that it is unseemly for him to scuffle with the hag, slices her neatly in two with a well-aimed throw of his knife, so that she looks like two tubs standing side by side. They find the great boar Twrch Trwyth busy devastating Ireland. They drive him across the Irish Sea to Dyfed and hunt him across South Wales and into Cornwall. They take the comb and scissors from his head and drive him into the sea. Finally they return to confront Ysbadadden. The giant agrees that his time has come and he consents to die. They cut off his head and set it on a stake and that night Culhwch sleeps with Olwen. Nothing more is said about the magic vessels.

The motif of death and regeneration suggested by the otherworld vessels seems to lie behind the central theme of the story. In terms of the seasons Ysbadadden is winter, barren as stone. His daughter is the promise of spring. When she loses her virginity, winter dies and new life blossoms in the summer.

There is a brief reference in the story to a myth in which the seasonal symbolism is unmistakable. Creiddylad, the most majestic maiden of the British Isles, ran away with Gwythyr son of Greidiawl (Victor son of Scorching), but before he had slept with her Gwynn son of Nudd carried her off by force. This happened in the north (the direction of winter and cold) and when Arthur heard of it he went north and made a treaty between Gwythyr and Gwynn. It was agreed that the girl should stay at home in her father's house, unmolested, and every year on May Day (which marked the beginning of summer in the Celtic calendar) Gwythyr and Gwynn should fight in single combat. This was to happen every year until the Day of Doom, and the one who was the victor on the Day of Doom should have the girl.

Without necessarily suggesting any direct influence, since myths of this kind are widespread, the story can be compared with the Greek myth of the rape of Persephone, the corn maiden. She was carried off by Hades, the god of the underworld and lord of death. When the whole earth became barren as a result, a compromise was arranged. Persephone was to spend part of each year on earth and part underground with Hades: hence the turning wheel of the year, the alternation of summer and winter, fertility and sterility, life and death. In the Welsh myth Creiddylad symbolizes the land, the crops and herds, the majesty and bounty of nature. Gwynn son of Nudd, the lord of the otherworld, and Gwythyr son of Greidiawl represent the rival powers of winter and summer, which every year contend for mastery. Arthur's role in the myth makes him responsible for establishing order in the world on a cosmic scale, as the real Arthur had done on the smaller stage of fifth-century Britain.

A similar seasonal theme probably lies behind a story in the *Life of Gildas* by Caradoc of Llancarfan. Gildas was at Glastonbury, the City of Glass, when Melwas, the King of the Summer Country, or Somerset, carried off Guinevere, Arthur's wife. He held her captive in Glastonbury, which was girdled by marshes and streams (as it was in fact in the real Arthur's time). Arthur searched for her for a year before he discovered where she was. He raised troops from Devon and Cornwall, and set siege to Glastonbury. Gildas and the Abbot of Glastonbury made peace and Guinevere was restored to Arthur.

The area round Glastonbury, the Isle of Glass or Isle of Avalon, was traditionally connected with the otherworld and Arthur here invades it to rescue his wife. Melwas seems to be playing the same role as Hades in the myth of Persephone or Gwynn in the myth of Creiddylad. He carries off a woman who represents the life of nature and the fertility of the land. In releasing her from captivity and restoring her to the human world, Arthur brings back from the otherworld a priceless treasure, the force which regenerates life.

Originally, it may have been a true story. Two of the Welsh triads refer to it, but in them Melwas is Medraut (Mordred) and the consequence of his rape of the queen is the battle of Camlann. This may preserve a genuine tradition of the events which led up to Arthur's last battle and death: a rebellion against him by Medraut, or Mordred, one of his relatives or henchmen, who abducted Arthur's wife. Certainly this became the standard medieval story of Arthur's downfall.

The triads are lists of people or objects or events grouped in threes: Three Red Ravagers of the Island of Britain, Three Fortunate Concealments of the Island of Britain, and so on. They were used as aids to memory by bards and story-tellers. One of them says that Arthur had three wives, all named Guinevere. The first Guinevere was the daughter of Cywryd Gwent. The second was the daughter of Gwythyr ap Greidiawl, the personification of summer in the myth of Creiddylad. The third was the daughter of a giant. Three was a magic number to the Celts, a number of completeness and sacredness. Three of anything made a complete set (and so Cuchulain had 'the gift of beauty, the gift of form, the gift of build' and 'the gift of battle, the gift of fighting, the gift of conflict'). Celtic mother goddesses and fertility goddesses were imagined in groups of three. It may be that the rape of Guinevere was a historical event which came to be treated as a nature myth, which in turn caused Guinevere to be ranked as a nature goddess and so led to the notion of the three Guineveres. An additional possibility, suggested by Irish mythology and custom, is that Arthur's wife came to be regarded as a personification of the land of Britain, the land which Arthur 'married' in his capacity as king.

There are stories about Arthur in several lives of Celtic saints. They invariably show the saint righteously rebuking Arthur and putting him in his place as a mere king who should be humbly subordinate to men of God. In Somerset, for instance, Arthur blasphemously tried to use the altar of St Carantoc's church as a table, but the saint forced him to repent. In Wales he coveted Bishop Paternus's tunic and tried to take it. The bishop said, 'Let the earth swallow him up,' which it promptly did, and Arthur was left stuck in the ground up to his chin until he begged for forgiveness. These tales belong to a genre of ecclesiastical propaganda, intended to portray Christian holy men as superior to the great hero of popular and semi-pagan belief.

That the propaganda was ineffective is shown by the tenacity with which people believed that one day Arthur would return. A poem in the Black Book of Carmarthen mentions various famous heroes and their burying-places, but of Arthur it says: 'An eternal wonder [*anoeth*] is the grave of Arthur.'[12] There is no way of being sure what this means, but it may refer to a belief that, though seriously wounded in his last battle at Camlann, the hero had not died but had been translated to a different plane.

This was certainly believed later. In 1113, when some French priests were travelling in Cornwall, there was an ugly scene when one of their servants contradicted a Cornishman who said that King Arthur had not died like other men but was still alive. William of Malmesbury, a few years later, said that no one knew where Arthur was buried and there were old prophecies that he would return. The Welsh at this time were said to believe that they would reconquer all Britain when Arthur came back to lead them. The Bretons, it was said, expected Arthur's reappearance as confidently as the Jews awaited the Messiah and anyone who ventured to tell Bretons that Arthur was dead would be lucky to escape being stoned.

Arthur was so important to the Welsh, the Cornish and the Bretons, as a symbol of their Celtic identity and a source of pride in it, that they could not allow him to be relegated to the past. He was not the only legendary figure who was treated in this way, though he was by far the most important. A Welsh propaganda poem of about 930, *Armes Prydein* (The Prophecy of Britain), predicts the imminent return to earth of Cynan and Cadwalader, two heroes of the past, who will lead the Welsh, Cornish, Bretons and Scots against the English, drive them back overseas to the lands from which they came and recover Britain for the British. The poem was written when just such a grand Celtic alliance was being concerted, though it came to nothing.

There were various different stories of what happened to Arthur after his last battle. According to the best-known one, he was spirited away to the otherworld Isle of Avalon, the paradise of apple trees in the west. There his wounds were healed and there he would enjoy peace and rest until the time came for him to return to Britain and lead his countrymen once more to victory over their enemies. The story is an example of the principle that the true hero must not stay in the otherworld but must return to his people, who need him.

If the tales of Arthur and his men had remained obscurely pent up in their original home on the north-western fringes of Europe we might now be left with little more than scattered folk beliefs about him. But in the eleventh and twelfth centuries Celtic legends were injected into the bloodstream of European culture by professional story-tellers. Travelling minstrels exported to England and the Continent stories which had been handed down for generations in Wales, Cornwall and Brittany (and which were often influenced by Irish legends). Besides the stories of Arthur and his men, this mass of Celtic material included tales of other heroes who were soon to be linked with Arthur.

The wandering minstrel earned his living by going from court to castle and fair to market, entertaining people with poetry and music, love stories and tales of adventure. If he was particularly good at it, he might settle down at one court and stay there for years. Listening to him was a favourite pastime in an age when there were few books and few who could read them in any case. Audiences already familiar with, and perhaps beginning to be bored by, the legends of Alexander the Great and the exploits of Charlemagne and his paladins, were now enthralled by Arthur and his warriors. Peter of Blois, a widely travelled civil servant of the twelfth century, said that skilful story-tellers in his day reduced listeners to tears with tales of Arthur, Gawain and Tristan.

Little is known about the story-tellers themselves. There was a Welsh minstrel named Bleddri whose Arthurian tales made a profound impression at the court of the Counts of Poitou early in the twelfth century. Other Welsh entertainers may have earned a living in England and France with stories of Arthur, but the leading role seems to have been taken by Bretons. In the fifth and sixth centuries many Britons had fled from the Saxons to their Celtic kinsfolk in Brittany. Close links of trade and communication connected Brittany with

Cornwall and Wales, and the Breton language was close to Cornish and Welsh. But Brittany was part of France and was inevitably influenced by French culture. Breton minstrels were ideally placed to translate Celtic traditional stories into French and modernize them for French-speaking audiences.

The most important single event in this diffusion of the Arthurian legends was probably the Norman conquest of England. William of Normandy's army at Hastings in 1066 included Breton contingents with their own minstrels. The Normans swiftly overran England and the Bretons were rewarded with grants of English land. They presumably brought their tales of Arthur back to his original homeland with them. In addition, England now had a governing class which spoke French and which was in direct contact with both France and Wales. The Normans drove into Wales and reduced large areas of it to an uneasy obedience. Norman lords in Wales and on the Welsh borders may well have come across stories of a great hero who fought against the ancestors of the now subjugated English. These stories may then have spread into Norman circles in England itself and soon across to France.

Anyway, the stories reached France by one route or another and by about 1100, apparently, they had spread as far as Italy. A stone-carving above the northern doorway of Modena Cathedral, believed to date from before 1120, shows knights on horseback attacking a moated fortress. Fortunately, the sculptor put in the names of the characters. The leading horseman is Artus de Bretania (Arthur of Britain), and among the others are Galvaginus (Gawain) and Che (Kay). Inside the fortress is a female figure labelled Winlogee. This name is half-way between the Breton and French forms of Guinevere and the carving evidently illustrates the story of the abduction of Guinevere and her rescue by Arthur and his men.

Although the minstrels' tales were designed to please, they did not lose their old mythological character as stories believed to be true. The modern concept of fiction had not yet been born and making a story up out of whole cloth was regarded as telling lies. This is why medieval audiences were upset if a story-teller's version of a tale differed too sharply from the version they had heard before. Minstrels had to ride their inspiration on a reasonably tight rein, and the more adventurous of them devised ingenious defences against accusations of inconsistency. One was that in King Arthur's day God had altered the scenery from time to time. When one of the heroes visited a country which he had explored before, God mercifully changed the castles and customs and perils he encountered, to preserve him from boredom.[13]

In practice, inevitably, traditional tales were altered as they were passed on by word of mouth from one generation to the next, but each story-teller worked within the convention that his story was a true story, not one which he had made up. However, to make the Arthurian tales seem real to a medieval audience their historical, social and psychological setting had to be brought up to date. The result was that the Arthur of Celtic tradition turned into a medieval king and his war-band became feudal knights. Legends imbued with primitive heroic values were gradually transformed into romances of chivalry.

A major landmark in this process is Geoffrey of Monmouth's *History of the Kings of Britain*, which came out in England in the 1130s. It contains the first full and connected account of Arthur which has survived, and probably the first ever written.

The Boar of Cornwall

Geoffrey of Monmouth's *History* was intended to fill a yawning gap by providing a written account of British history from the earliest times to the Saxon conquest of England. Its author, apparently, grew up on the Welsh borders and was himself of Breton or Welsh descent. He drew his book partly from previous writers, partly from current oral traditions and partly from his own fertile imagination. The bulk of the book is fiction, but since it was accepted as authentic history by everyone except a few crabbed scholars for four hundred years, it had considerable influence. It was immensely popular and by giving Arthur a story with a beginning, a middle and an end, it provided a rough framework to which later authors could attach more stories of the great king and his court.

According to the *History*, Arthur's distant ancestors came from illustrious Trojan stock. He was descended from no less a person than Aeneas, Prince of Troy, the hero of Virgil's *Aeneid* and the legendary forefather of the Caesars. After the siege of Troy, when the Greeks took and sacked the city, Aeneas escaped to Italy. He had a great-grandson, Brutus, who recruited a following among other Trojan expatriates and led them to a rich and beautiful island in the northern seas. It was called Albion, and was inhabited only by giants. The Trojans drove the giants away into the mountains, settled down in the country and built the city of New Troy, subsequently renamed London, as their capital. Brutus ruled the island, which was called Britain after him, and all the later kings of the Britons were descended from him.

This was an old story and the bones of it appear in Irish annals of the seventh century and in Nennius. Albion was a genuine early name for Britain. The legend put the British ruling house on a level with other European dynasties which also claimed a Trojan lineage. (Descent from Aeneas, incidentally, gives Arthur a divine ancestry, for in classical mythology Aeneas was the son of Venus, the beautiful goddess of love.)

From Brutus to Arthur in the *History* stretches a long line of legendary kings, including King Lear, King Cymbeline and King Coel, better known now as Old King Cole. We reach the dawn of the Arthurian age in the time of King Constantine II. This 'king' was in reality a Roman officer who was proclaimed emperor by his troops in 406 and took his army across the Channel to fight the barbarians in France. Geoffrey's *History* gives him three sons: Constans, who succeeded him; Aurelius Ambrosius (the Ambrosius Aurelianus of Gildas); and Uther. The young Constans was murdered by the wicked Vortigern, who usurped the throne and brought the Saxons to England as mercenaries. Aurelius and Uther, who were still children, were hurried over to Brittany for safety by their faithful attendants.

At this point the *History* introduces a new character, whose fame later rivalled that of Arthur himself. When the Saxons turned against Vortigern, the usurper fled in terror to Wales. There he encountered a mysterious boy from Carmarthen, whose father was a demon and whose mother was a princess, and who was gifted with strange and formidable powers, including the ability to see into the future. His name was Merlin. He told Vortigern that the conflict between the Britons and the Saxons would go on until the coming of the Boar of Cornwall. The Boar would trample the Saxons under his feet and conquer France and the islands of the sea, and his end would be shrouded in mystery.

The boar is the 'sovereign beast', renowned for savagery, strength and indomitable courage, the most dangerous of all the beasts of the chase. The Boar of Cornwall is Arthur and the great hero is here a figure of destiny, whose coming was foretold by a prophet and whose career was planned by unseen forces, guiding events from behind the scenes.

Aurelius and Uther invaded England from Brittany, defeated the Saxons and burned Vortigern alive in his own stronghold. Aurelius recovered his rightful throne, though the numerous Saxons still in England remained an ever-present threat. To advise him, Aurelius summoned Merlin to his side. It was Merlin who succeeded, with extraordinary engineering skill, in bringing Stonehenge to England from Ireland as a monument to Britons who had been massacred by the Saxons. The colossal stones of the Giants' Ring or Giants' Dance, as it was called, originally came from Africa and had been taken to Ireland by giants far back in the past. Guarded by Uther and an army, Merlin had the Giants' Ring carefully dismantled, stone by stone, brought across to England by sea and reassembled on Salisbury Plain. This story increased the prestige of Arthur's prophet, and so indirectly of Arthur himself. (As we shall see later, it is not quite so far from historical truth as it looks.)

Aurelius was poisoned by a treacherous Saxon and was buried at Stonehenge. He was succeeded by his younger brother Uther, known as Pendragon – chief dragon or dragon's head – because a huge star in the form of a fiery dragon was seen in the sky at the moment of his accession. Uther was opposed by the Saxons in force. He roundly defeated them, restored peace one more and held a great celebratory feast in London.

At the feast Uther was struck with passionate desire for Ygraine, the most beautiful woman in Britain. Unfortunately, she was the wife of Gorlois of Cornwall, one of Uther's most loyal and powerful supporters. The king's interest in her was so obvious that Gorlois hurried her back to Cornwall and refused to return to court when summoned. The furious Uther gathered an army and invaded Cornwall. Gorlois shut his wife up in the impregnable fortress of Tintagel on the northern Cornish coast, while he himself remained with his troops. Tormented by his longing for Ygraine, Uther consulted Merlin. Merlin gave him magic drugs which made him look exactly like Gorlois. In this disguise Uther went to Tintagel by night and had no difficulty in entering both the castle and Ygraine's bed. That night she conceived Arthur.

Arthur's Family Tree

According to Geoffrey of Monmouth

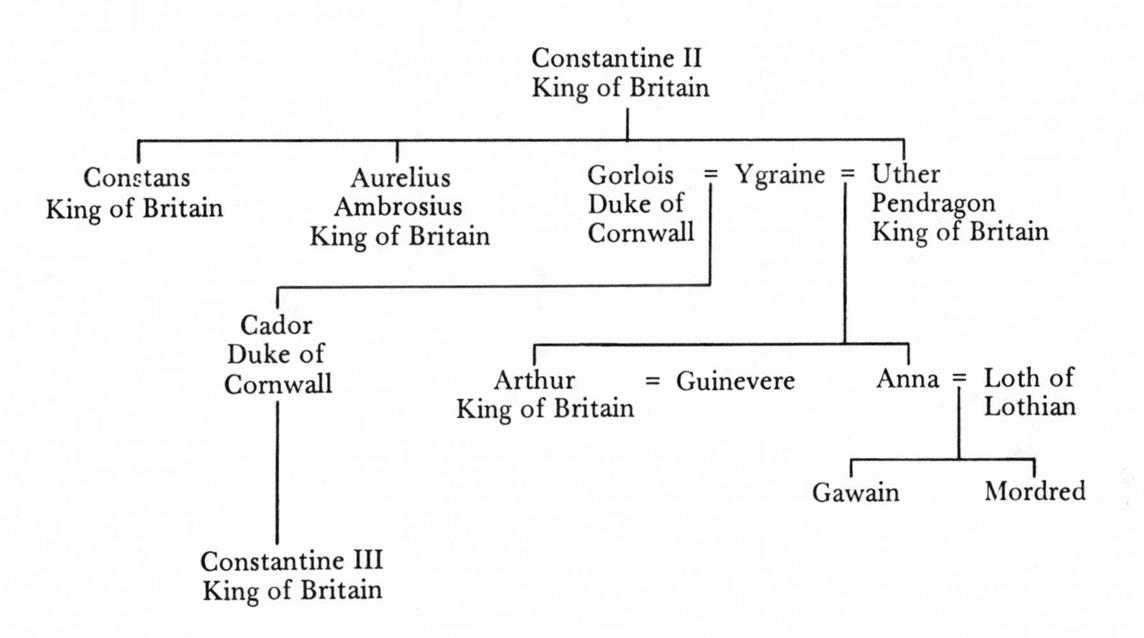

That same night Gorlois led an attack on Uther's army and was killed. Changed back into his normal form, Uther seized Ygraine and, apparently, they were promptly married. Besides Arthur, she later bore him a daughter, Anna, who married a British war-leader, Loth of Lothian. Loth and Anna had two sons, Gawain and Mordred.

Uther fell victim to a Saxon plot and was poisoned, like his brother, and he too was buried at Stonehenge. Arthur was now fifteen and already outstanding for courage, generosity and an innate goodness for which people loved him. The leading Britons insisted that he be crowned king at once, to lead them against the menacing Saxons.

This legend makes Arthur King of Britain by the respectable claim of hereditary right. It links him with Ambrosius Aurelianus, who led the British resistance to the Saxons before Arthur, and Uther Pendragon is also neatly fitted in. Uther's antecedents are uncertain, but his name may have been derived originally from a misunderstanding of the Welsh phrase *Arthur mab uthr*, 'Arthur the terrible', as Arthur son of Uther. This suggests that there was a mystery of some kind about Arthur's father.

Later writers added other incidents to Arthur's early career, but Geoffrey of Monmouth's story of his conception at Tintagel remained unaltered in all essentials throughout the history of the legends. A story which weathers time so successfully must have a profound psychological appeal. Geoffrey probably based it on established tradition. Arthur is closely connected with Cornwall in 'Culhwch and Olwen' and the Welsh triads. There are traces of a Welsh story that Arthur's mother was Eigr (Ygraine) and that on her side he came from the royal house of Dumnonia, a British kingdom covering modern Cornwall, Devon and northern Somerset. The simplest explanation of this tradition, though it is quite impossible to prove it, is that the real Arthur actually was related to the Dumnonian kings. Tintagel may have had a local legend that Arthur was conceived there, which Geoffrey adopted, or he may have located the story there because he thought it a suitably impressive and romantic setting, as indeed it is. It seems likely that Geoffrey himself added Merlin to the story.

However it grew up, the story gave Arthur's birth an air of magic and mystery in symmetry with the mystery surrounding his death. The classical myth of Jupiter and Alcmene might have been an influence on it. Jupiter succeeded in sleeping with Alcmene by taking the form of her husband. Her son by the god was the great hero Hercules. Another possible source is the legend of Alexander the Great. An Egyptian king named Nectanebus, who was skilled in magic, was driven out of his country and took refuge in Macedon, where he told Queen Olympias that the god Ammon desired her and would father a child on her. Nectanebus himself, disguised as the god, duly did so, and the child was Alexander. Nectanebus was said to have made love to Olympias in the form of a dragon, which interestingly recalls Uther's title of Pendragon.

There is a parallel closer to home in an Irish story about King Mongan, though it may be later in date than the original legend of Arthur's conception. Mongan was a real man, the ruler of a petty kingdom in Ulster in the seventh century. His putative father was King Fiachna, but according to the story his real father was the sea god Manannan mac Lir. Fiachna was fighting in Scotland and found himself close to defeat when Manannan suddenly appeared and offered him victory. The condition was that Manannan be allowed to sleep with Fiachna's wife. When Fiachna agreed, Manannan went to Ulster, took the form of Fiachna and so deceived the queen. Their son was Mongan. When the child was three nights old, Manannan came for him and took him away to the otherworld, where he was brought up until he was a young man. Something similar happens in the later legend of Arthur.

Behind the legend of Arthur's conception there may lie an older tale that he was sired by a god in disguise. This would suit the semi-divine status which he was given in Celtic tradition. In legend the birth of a hero is usually surrounded by an uncanny, supernatural atmosphere, and it would be surprising if no such story had been told about Arthur. In many cases the hero's begetting is in some way against nature or contrary to accepted morality and custom. His father is a god or an animal or a stranger from another race; or the hero is the child of incest, rape, trickery or magic. This is true of holy men as well as secular heroes. According to their legends, many Celtic saints were conceived in questionable circumstances. 'St David and St Cynog were the products of rape, St Cadoc's mother was abducted and St Lonan's mother was tricked into having intercourse with a man other than her lover. St Cenydd and St Cuimine Foda were born of incest . . . St Finan's mother was impregnated by a red-gold salmon when she was bathing in Loch Lein.'[14]

The simple explanation is that a hero of superhuman powers cannot appropriately be conceived in a normal coupling of man and wife, but the logic of these stories goes far beyond this. The hero is born of an act which is perverse and wrong, which outrages nature and morality. The twistedness of the act and the violence of the passion which inspires it create or liberate a powerful magic force, with which the hero is imbued. It not merely enables him to transcend human limitations, but positively drives him to do so. He is possessed by an overwhelming energy which smashes down all normal barriers. Here, probably, is the root of the legend's hold on the mind.

In the *History*, Uther is seized by ravenous and overpowering desire when he first sees Ygraine, as if by some force from outside himself. He can think of no one and nothing else. He fears he will die if his passion for her is not slaked and it drives him to rape her by magic, betray a loyal friend and endanger his throne. The implication is that the irresistible tidal wave of desire in which Arthur was conceived made him the hero he was.

As soon as he was crowned, Arthur moved against the Saxons and defeated them in a campaign which culminated in a decisive battle at Bath, in Somerset. This is evidently the battle of Badon Hill. He next subdued the Picts and Scots, and went on to conquer Ireland and Iceland. He married Guinevere, a girl of noble Roman-British family, brought up in Cornwall and, like Arthur's mother before her, the most beautiful woman in Britain. Encouraged by the fear which he inspired abroad, Arthur planned the conquest of all Europe. He made a start by subjugating Norway and Denmark. Then he invaded France and after nine years of campaigning he reduced the whole country to obedience.

Arthur's triumph was marred by the arrival of a terse message from Rome, ordering him to acknowledge Roman overlordship and pay tribute. His reply was to gather an enormous army from all his dependencies under his standard of the Golden Dragon. Leaving Guinevere and his nephew Mordred in charge in Britain, Arthur sailed for France. The Romans mustered an even larger army, with levies from Spain, Africa and the East, and the two armies met in Burgundy. After a tremendous battle of thrust and counter-thrust, in which Arthur and Gawain fought like men possessed, the Romans were driven from the field. On the British side, Bedivere was killed and Kay was mortally wounded.

Arthur was preparing to invade Italy and bring Rome herself to her knees when news came that he had been betrayed in Britain. Mordred had seduced Guinevere and seized the throne. Arthur hurried back to England, where the treacherous Mordred had compounded his villainy by hiring Saxon mercenaries and gaining support from the Picts and Scots. At the first encounter with Mordred's troops, Gawain was killed. Mordred retreated to Winchester and Guinevere, in despair, fled to a nunnery in Caerleon.

Arthur drove Mordred back into Cornwall and to the final battle on the river Camel. Mordred's army was routed and Mordred himself was killed in a charge led by Arthur in person, but tragically, in this last fight against barbarism, Arthur fell. 'Arthur himself, our renowned king, was mortally wounded and was carried off to the Isle of Avalon, so that his wounds might be attended to. He handed the crown over to his cousin Constantine, the son of Cador Duke of Cornwall: this in the year 542 after our Lord's incarnation.'[15]

Merlin's prophecy about the Boar of Cornwall has now been fulfilled. Arthur has crushed the Saxons and conquered France and his end, like his beginning, is shrouded in mystery. The *History*'s account of the king's last days is hedged with caution. The story of Mordred's treachery may have been founded on fact, but in dealing with it Geoffrey of Monmouth turns suddenly coy. He may merely have wished not to dwell on a shameful episode which dishonoured Arthur, or was he perhaps aware of a mythological implication? It seems likely that in Welsh tradition Guinevere had come to represent the land of Britain, which Arthur 'married' as king. The conclusion could be drawn that Mordred's seizure of the throne was related to his sexual conquest of the queen, whose union with Arthur had been barren. A writer hoping for preferment in the Church might decide not to explore this pagan theme. With similar caution, he has the dying Arthur taken to Avalon but says nothing about the belief in his return. The Norman masters of England did not welcome hints of a Celtic reconquest.

Geoffrey of Monmouth was writing as a historian, or a pseudo-historian, and his Arthur is consequently less of a superhuman and supernatural figure than before. The Welsh picture of Arthur as a great conqueror has now been developed to the point where he is almost a British Charlemagne. He is an idealized medieval ruler, a skilful general and a ferocious fighter, proud, impetuous, ardent and inspiring. He is haughty to the proud but kind to the humble. He is a faithful but not tiresomely pious son of the Church and he keeps a splendid court. Brave men are drawn to him as before, but they are feudal knights, not beings of Celtic legend half-way between men and gods. There is one episode in which Arthur kills a giant, but for most of the time he has been brought back from a semi-supernatural plane to the natural one. He and his knights fight men, not monsters.

Robert Wace's French version of the *History*, the *Roman de Brut*, gives an even more vivid impression of the splendour, brilliance and elegance of Arthur's court, the beauty of its women and the valour of its knights. The great king moves in a golden glow of heroism and magnificence. He becomes a more human figure, who grieves for Gawain and other knights killed in his service. Arthur's love for Guinevere is emphasized, and so is their failure to produce an heir to the throne, which for medieval audiences was a tragedy for king and kingdom.

Wace mentions the belief that Arthur was still alive in Avalon and would reappear one day, but without committing himself to it. 'Men have ever doubted, and – as I am persuaded – will always doubt whether he liveth or is dead.' A far less sophisticated version of the story in English, Layamon's *Brut*, is more positive. When Arthur is wounded in the final battle in Cornwall, he says: 'And I will fare to Avalun, to the fairest of all maidens, Argante the queen, an elf most fair, and she shall make my wounds all sound; make me all whole with healing draughts. And afterwards I will come again to my kingdom, and dwell with the Britons with mickle joy.' Then a boat is seen floating in over the sea, with two women in it, 'wondrously formed'. They take Arthur into the boat and away across the waves.[16]

This is the earliest surviving story of fairy women coming to spirit the dying king away in a boat, but there were already tales that he had gone to an otherworld island to be healed by a queen, whose name is usually given as Morgan. Geoffrey of Monmouth's *Life of Merlin* (written fifteen years or so after his *History*) records a story that Merlin and Taliesin took the wounded king to the Isle of Apples or Fortunate Isle, where the beautiful Morgan lived with her eight sisters. Morgan said she could heal him if he stayed with her for a long time. Much relieved, Merlin and Taliesin left Arthur with her and sailed home. A sarcastic poem, *Draco Normannicus*, has Arthur taken to the Isle of Avalon to be healed and made immortal by

Morgan, who is his sister, and says that he is now ruling over the Antipodes, at the opposite side of the world.

Writers in the twelfth century, rightly or wrongly, derived the name Avalon from the Welsh word for an apple, *afal* or *aval*. In Celtic mythology the apple is connected with immortality and is the characteristic fruit of the otherworld, where there is no death. It is also the fruit of female sexuality: the reason for this can be seen by cutting an apple in half. There were Irish stories of otherworld islands rich in apple trees and inhabited by beautiful and amorous women, and the story of Arthur going to Avalon to live immortally with Morgan belongs to this genre.

At some stage Avalon was linked with Glastonbury. There was evidently an old tradition that Glastonbury, which was ringed by marshes, was an otherworld island. According to Caradoc of Llancarfan, the British name for Glastonbury was Ynis Gutrin, which meant Isle of Glass, hence the 'Glas' in Glastonbury. An island or fortress of glass was one of the forms of the Celtic otherworld, related apparently to the belief that the otherworld had walls of air, which were transparent but impenetrable. Late in the twelfth century, Gerald of Wales, a Norman historian who was born in Pembrokeshire and spent many years in Wales, said that Glastonbury had been known to the Welsh from time immemorial as Avalon, the Isle of Apples, because apples grew there in abundance.

A few years earlier the monks of Glastonbury Abbey had created a sensation by announcing that they had discovered the bodies of Arthur and Guinevere, buried in the abbey grounds. In 1184 the abbey had been badly damaged by fire. Rebuilding began on a lavish scale with the help of funds provided by King Henry II. Henry died in 1189 and his successor, Richard I, cut off the supply of money, which left the monks in serious difficulties. In 1191 they claimed to have found Arthur's grave.

Gerald of Wales went to Glastonbury to investigate, a year or two later. He fiercely disapproved of the popular belief that Arthur would return, which he regarded as ridiculous and politically dangerous. He insisted that Avalon was merely an old name for Glastonbury and that Morgan had been a respectable Roman-British matron, a relative of Arthur, who tended the dying king when he was taken to Glastonbury after his last battle. He was consequently delighted by the discovery of Arthur's grave, which seemed to prove that the hero was quite definitely dead. His welcoming attitude may have numbed his critical faculties.

According to Gerald, an old Welsh minstrel had told Henry II where the body of Arthur was buried, sixteen feet deep, in the cemetery at Glastonbury. Henry passed on the information to the monks, who dug as instructed and found the skeletons of a man and a woman, lying in the hollowed-out trunk of an oak tree. There was also a lock of woman's hair, which still looked fresh and golden, but when one of the monks picked it up it crumbled to dust. The man's skull and thigh-bones were prodigiously large. On the skull were the marks of ten wounds, including one great cleft which had apparently been the cause of death. In the grave was a lead cross, with the inscription: *Hic jacet sepultus inclitus rex Arthurus cum Wenneveria uxore sua secunda in insula Avallonia* ('Here lies buried the famous King Arthur with Guinevere his second wife in the Isle of Avalon').[17]

The bodies were reverently removed from the grave and were later reinterred in a marble tomb in the abbey church. The lead cross, or a lead cross, survived into the eighteenth century. William Camden, the antiquary, published a picture of it in 1607. The inscription, in lettering which has been very tentatively dated to the tenth or eleventh century, reads: *Hic jacet sepultus inclitus rex Arturius in insula Avallonia*, with no mention of Guinevere. The earlier reference to Guinevere as Arthur's second wife is puzzling. It may be connected with the Welsh tradition that he had three wives, all named Guinevere.

The monks certainly did dig a deep hole in their cemetery, because traces of it were found during archaeological excavations at Glastonbury in 1962. The question is, what did they find in it? One answer is that the whole discovery was a pious fraud, designed to attract pilgrims to Arthur's grave and so bring the abbey prestige and the funds which it badly needed; and possibly encouraged by the English government in the hope of putting paid to politically undesirable talk of Arthur's return. The problem here, however, is the hollow tree. If the monks salted their dig with a skeleton they intended to pass off as Arthur's, it seems improbable that they would have 'found' it in such an eccentric receptacle.

A second possible explanation is that the real Arthur and his (second?) wife actually were buried at Glastonbury. The burial was hushed up for some reason, but knowledge of it was secretly preserved at Glastonbury and in Welsh bardic circles. The lead cross was put in the grave, to identify it, about 950, when it is known that the level of the old cemetery was raised. In this case the 'hollow tree' might have been a canoe, of the type used in the Glastonbury marshes. If Arthur was buried in a canoe, this could have contributed to the growth of the legend that the dying king had been borne away to Avalon by boat.

This explanation is unlikely. If the location of Arthur's grave was known at Glastonbury all along, it is hard to see why this information, which was of political value to the kings of England, was not disclosed before. That it was revealed when the abbey was in serious need of money makes it suspect. Another difficulty is that sixty years or so before the discovery, William of Malmesbury, who was interested in Arthur, went to Glastonbury and carefully researched a history of the abbey, in which he says nothing whatever about Arthur being buried there. Another problem is the discrepancy between the inscriptions on the lead cross. Either Gerald of Wales misremembered the inscription or the cross which he saw was not the one which Camden saw.

A third possible explanation is that the monks discovered a genuine grave, but not Arthur's grave. They may have found a burial of about 1500 BC, when chieftains were sometimes interred in boats or canoes. The unusual size of the skeleton and the marks on the skull showed that these were the remains of a formidable warrior. The monks decided that the grave was probably Arthur's, that it would redound to the honour and profit of the abbey if it was, and that it would be a godly work to help the identification along by faking the lead cross. This seems the most likely solution, and yet one cannot help hoping that the monks really did find Arthur's grave and that one of them did touch, for a moment, a lock of Guinevere's bright hair.

Anyway, the discovery was widely accepted as genuine at the time and it reinforced Glastonbury's reputation as a place of ancient and mysterious sacredness by firmly identifying it as Avalon. What the discovery did not do was to extinguish the belief that Arthur was still alive and would return. There were stories of him living in a magnificent fairy palace inside Mount Etna in Sicily, in a hollow hill in England, near Mont du Chat in Savoy or in the depths of the sea; of Arthur and Gawain living happily at Morgan's court on an island in the Mediterranean; of Arthur and Morgan dwelling in a mysterious paradise in the Far East, in India or on the eastern shore of the

Red Sea. Arthur was said to ride through the sky or the forest on wild nights with his huntsmen and his hounds, and the cries of the hunt and the baying of the hounds could be heard in the howling of the wind. He flew about the world in the form of a raven or a crow, or he and his court lay sleeping in a cave deep beneath a hill waiting for the time to wake and return to their kingdom.

Most writers were too sophisticated to accept these stories but the belief in Arthur's return survived at popular levels for centuries. It was still alive in England in Malory's time and traces of it lingered on in Welsh, English and Breton lore into the nineteenth century. The persistence of the belief shows how strongly the legend of the great king had gripped the imagination of ordinary people. The Arthur who crushed the Saxons had come to stand for a golden age in the past, when right was upheld and wrong put down, when might and right were one. The hope that this golden time would come again lasted for well over a thousand years after the bones of the real Arthur had crumbled into dust.

2

Arthur and the Round Table

'By my head,' said Sir Lancelot, 'he is a noble knight and a mighty man and well-breathed; and if he were assayed,' said Sir Lancelot, 'I would deem he were good enough for any knight that beareth the life. And he is gentle, courteous and right bounteous, meek and mild, and in him is no manner of mal engin, but plain, faithful and true.'

Malory, *Le Morte Darthur*

When the Celtic tales of Arthur and his men entered the mainstream of European culture, they turned into stories about a feudal king and his knights. Feudalism had developed originally to provide security in a chaotic world. Massive German migrations across the Rhine in the fourth and fifth centuries destroyed the Roman Empire in the West and replaced it with comparatively small and unstable German kingdoms, which in their turn were fiercely harried from the outside by Arabs, Slavs, Magyars and Vikings. Charlemagne briefly recreated the western Empire (excluding Britain) in the eighth century and, like Arthur before him, swiftly became a legendary hero. For most of the time, however, the West lacked any strong central authority with a standing army, like the old Roman army.

The result was that power passed to local lords, who kept a rough and ready peace in their own areas, sufficient to allow their peasants to raise crops, cattle and the next generation of farmers. Each lord held his land from a higher lord, who protected him on condition that he fought for the higher lord when required. The higher lord was similarly the vassal of a higher lord still, and so on up the pyramid to the king, who could call out all the lords and their men as a national army. This was the system in theory at least, though in practice it was far more complicated.

The feudal system had its roots in German tribal society, which closely resembled Celtic tribal society. All three systems were dominated by warrior elites with similar values and attitudes, which made it easier for heroic Celtic legends to take on a feudal colouring and attract the aristocracy of the middle ages. The curious thing is that the Arthurian legends flowered in a western Europe whose ruling class was German in origin, not Celtic, while native German myths and legends had far less impact. Celtic legends had greater force, perhaps because Celtic paganism had resisted Christianity more effectively and retained its vitality for longer than German paganism. And the reason for this may be that unlike the conquering Germans and their descendants, the Celts felt themselves to be a disadvantaged minority. Though converted to Christianity, they clung more tenaciously to their pagan heritage, as they have continued to do ever since.

The backbone of the feudal system was the career heavy cavalryman, *chevalier* or 'horseman' in French, knight in English. He was a professional soldier, trained to arms from boyhood. He might live in his lord's hall, but more often he lived on his fief, the estate which he held in return for his military service. He might have only a small estate or he might be a great lord with extensive acres and many lesser knights at his call. Whatever his status in the ranks of knighthood, compared to everyone else he was an aristocrat. He was one of the *bellatores*, who fought, as distinguished from the *oratores* or priests, who prayed, and the *laboratores*, the great majority of the population, who worked. Through their control of the land, the knights were the ruling class. They ran their estates, kept order and administered justice.

Chivalry, the creed of the horseman, was the knight's social, moral and religious code. Its fundamental ideals were courage, loyalty, generosity and honour – the ideals of the primitive war-band, now embedded in a more complex and formal structure. Courage is naturally the first quality required of a fighter. Loyalty and generosity were vital because the feudal system was a network of mutual personal obligations, of vassal to lord and lord to vassal. Treachery was perhaps the most despicable sin which a knight could commit.

The key word in the vocabulary of chivalry was honour, which summed up all that was due to a knight and expected of a knight, and which consequently had a vast range of connotations and implications. It meant being held in respect, it meant behaving in a way which earned respect, and it meant paying proper respect to others.

It consequently implied a dread of being shamed, and shame rather than guilt was the code's principal sanction. Public approval was valued more highly than a clear conscience.

Honour meant straightforwardness, devotion to duty and reluctance to take unfair advantage of an opponent. It also meant not doing menial work or engaging in trade. Honour implied pride, and a knight would not tolerate an insult or accept a blow without returning it. In turn he behaved courteously to others. Honour meant protecting women, children, the weak and defenceless. It meant being a faithful Christian, protecting the Church and fighting her enemies. Pious authors put this last obligation first. Less pious authors, like Malory, sometimes omitted it altogether.

A knight's 'word of honour' was the most binding undertaking he could give. In stories, a knight giving his word to do something difficult, rash or apparently impossible, and then trying to keep his word, is a frequent springboard of the plot. A dramatic example of what it could mean in real life occurred when King John II of France was taken prisoner by the English at the battle of Poitiers in 1356. Carried off to England, he returned to France four years later to try to raise the enormous ransom which the English demanded. His son, Louis of Anjou, took his place in London as a hostage. Presently Louis escaped from England and refused to go back. Regarding this as an intolerable stain on his honour, John returned voluntarily to captivity in England, where he was greeted with profound admiration. His French advisers, who had now to raise the ransom, were less enthusiastic.

It has frequently been pointed out, sometimes with righteous indignation or smug satisfaction, that the ideals of chivalry were more honoured in the breach than in the observance. This is true, but we in our time are in no position to criticize past ages for brutality, treachery, rapacity and hypocrisy. Like other high ideals, the code of chivalry was not easy to live up to.

In the twelfth century, when the first Arthurian romances were written, French culture, permeated by the ideals of chivalry, dominated western Europe. It had an outpost in England, where the ruling class thought and spoke in French. In a period of comparative peace and security, with greater prosperity and leisure, the code of chivalry was tempered by softer, more civilized influences than had gone to its forging. Courage and fighting prowess were still essential qualities of the knight, but graceful manners, elegance, wit and refined taste were now demanded of him. What was more, he was required to be in love. The troubadours, the lyric poets of southern France, had brought romantic love and the idolization of women into fashion. The troubadours' prescriptions had to be heavily watered down before many people would swallow them, but the French were already establishing their reputation as masters of the civilized art of love. These gentler influences played on the new French literary genre of Arthurian romance, though the bedrock of the Matter of Britain remained the older heroic values of courage, loyalty and honour.

The new genre has a far more sophisticated flavour than the older stories. Its pioneers were such writers as Chrétien de Troyes in France and Hartman von Aue in Germany, who wrote stories about individual knights of the Round Table. Chrétien also wrote an unfinished story about the Grail, the earliest that has survived. These tales about the adventures of knights, called 'romances', were often based on Celtic tales and traditions, but medieval authors did not merely copy Celtic stories down. They altered, enriched and transformed them in the light of their own inspiration and experience. Though they did not consider themselves writers of fiction, they explored subtleties of character and plot, they investigated the psychology of their heroes and heroines. Their stories are nearer to modern novels than are the simpler Celtic legends.

The first half of the thirteenth century was the most productive period in the history of Arthurian literature. In France four sequels to Chrétien's unfinished Grail romance appeared, plus another important Grail story, *Perlesvaus*. In Germany Wolfram von Eschenbach completed his Grail epic, *Parzival*, by about 1210 and Gottfried von Strassburg's *Tristan* came out at about the same time. The main French romance of Tristan, or Tristram, was written about 1230.

Meanwhile, the next major step was being taken by writers who attempted to put on paper the entire history of Arthur, the Round Table and the Grail in a complete, coherent chronicle of the Matter of Britain. This involved the difficult task of fitting the stories and traditions together in a logical order, elucidating causes, motives and consequences, and filling in gaps. It was like writing a cross between an encyclopedia and a novel, and it has often been compared to the creation of a complicated tapestry of interweaving threads.

A Burgundian knight named Robert de Boron set out to produce a compilation of this kind, but apparently never finished it. The first and most influential complete account of the Matter of Britain was put together by unknown writers in France between about 1215 and 1230. It is known today as the Vulgate Cycle, because of its widespread acceptance and semi-canonical position in Arthurian literature. Another attempt to create a complete Matter of Britain was made in the 1230s. It is now known as the Post-Vulgate Romance, and not all of it has survived. One part which has is the *Suite du Merlin* (Sequel of Merlin).

Many more Arthurian romances remained to be written, and between 1250 and 1450 stories of the Round Table appeared in almost every western European language. The finest monument to the Matter of Britain, however, was still to be raised. In 1469 or 1470, in Newgate Prison in London, Sir Thomas Malory wrote the last words of 'the whole book of King Arthur and of his noble knights of the Round Table'. It was printed in 1485 by William Caxton and the title by which it is usually known, *Le Morte Darthur*, is Caxton's, not Malory's. It is based on earlier French and English sources. There are many confusions and inconsistencies in Malory's handling of his sources, but the result glows with the splendour and charm of a prose style and a heroic idealism that have captivated readers ever since. (For more details of Arthurian writers and books, see Appendix I.)

The Sword in the Stone

The outline of Arthur's career provided by Geoffrey of Monmouth and Wace was altered and amplified by later writers, who added numerous incidents to the great king's story. The standard legend of Arthur's early life comes from Robert de Boron's *Merlin* and its sequels in the Vulgate *Merlin* and the *Suite du Merlin*, which Malory followed. Arthur's father was Uther Pendragon, King of Britain. His mother was the beautiful Ygraine of Cornwall and he was conceived at Tintagel when Merlin, by magic art, enabled Uther to sleep with Ygraine in the form of her husband. That same night Ygraine's husband was killed, and soon after this Uther and Ygraine were married.

So far we are on familiar ground, but at this point the story diverges from the earlier track. For one thing, instead of disappearing from view when Arthur is conceived, Merlin plays an important part in the young hero's early career.

As soon as Ygraine's child by Uther was born, Merlin claimed him, chose the name Arthur for him and had him christened. He then sent the baby away to be brought up secretly in the country by a knight named Ector, with his wife and their young son Kay. A tradition grew up in Wales that Arthur spent his boyhood at Caer Gai (Kay's Fort), a hill fort near Bala in the remote mountains of Snowdonia. It was said that Kay spoke with a stammer all his life because the baby Arthur had taken his place at his mother's breast.

Arthur was kept in ignorance of his true parentage; he thought he was Ector's younger son. He consequently did not know that he had three older half-sisters, Ygraine's daughters by her first husband, the Duke of Cornwall. The eldest sister, Morgause, was married to King Loth of Orkney and was the mother of Gawain. The youngest sister was Morgan le Fay, the great enchantress. Both of them were to play sinister roles in Arthur's life.

Morgan le Fay is the same Morgan who, at the end of Arthur's life on earth, takes him away to Avalon for the healing of his wounds. Fay is an old word for fairy, but 'fairy' nowadays has all the wrong connotations. It suggests a little, insipid, sugary, gauzy-winged sprite out of a story for small children. There is nothing childish or insipid about Morgan and the other Arthurian fays, who are tall, commanding and seductively beautiful. Dominating, ruthless, sensual and unpredictable, they are sometimes benevolent and sometimes cruel. They have formidable magical powers and are intensely dangerous to cross. They are of a race older than man, and they are either immortal or live far longer than any human span. They appear in the human world whenever they wish, but their true home is in the otherworld or land of faerie. Although the legend makes Morgan Arthur's half-sister, she retains her faery characteristics.

Arthur's father, Uther Pendragon, died two years after Arthur's birth. There was a long, uneasy interregnum. No one except Merlin and Ector knew who the true heir was, or where he was. At last, when Arthur was almost fifteen, Merlin summoned the nobility of the realm to London at Christmas time, promising them a divine revelation of the identity of the rightful king. When everyone was in church hearing Mass, there appeared in the churchyard, in Malory's words: 'a great stone four square, like unto a marble stone, and in midst thereof was like an anvil of steel a foot on high, and therein stuck a fair sword naked by the point, and letters there were written in gold about the

Arthur's Family Tree

According to Malory

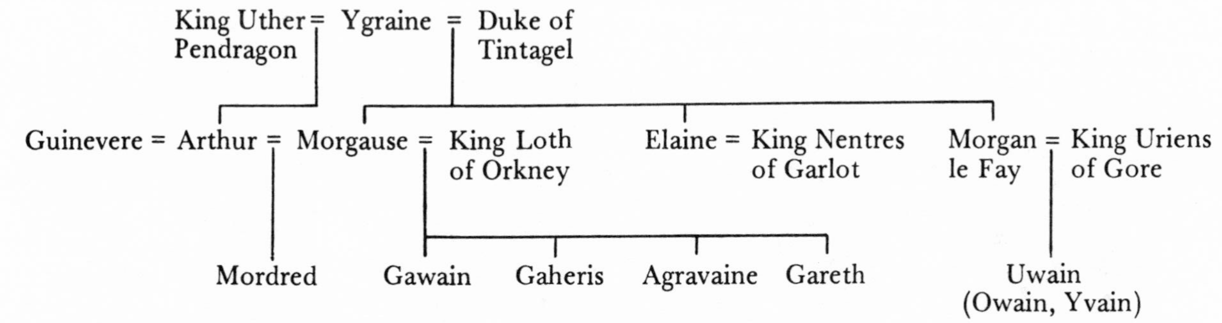

sword that said thus: WHOSO PULLETH OUT THIS SWORD OF THIS STONE AND ANVIL IS RIGHTWISE KING BORN OF ALL ENGLAND.'[1]

Many lords and knights crowded to the stone and tried to draw the sword, but could not budge it an inch. The youthful Arthur had come to London with Ector and Kay. When no one else was about, he went up to the stone and casually pulled out the sword. He naïvely gave it to Kay, who tried to pretend that he had drawn it from the stone himself. Ector quickly unmasked this deception. The sword was put back in place. Kay tugged at it in vain, but Arthur drew it out easily. Ector told Arthur that he had proved himself the rightful king and explained that he, Ector, was not Arthur's real father. Arthur was astonished and deeply upset: 'Then Arthur made great dole when he understood that Ector was not his father, for he said: "you are the man in the world that I am most beholding to, and my good lady and mother your wife that as well as her own hath fostered me and kept. And if ever it be God's will that I be king as you say, you shall desire of me what I may do and I shall not fail you. God forbid I should fail you." '[2]

Arthur was acclaimed king, to the fury of Loth and other great lords, who would not consent to be ruled by a mere boy of base blood. Merlin revealed the secret of Arthur's birth to them, but many of them did not believe him and others argued that if it was true, Arthur was a bastard. It was to take Arthur several years of hard fighting to bring them to heel.

This is a far more exciting and romantic account of how Arthur came into his kingdom than the earlier story. The struggles of Philip Augustus, who succeeded to the throne of France in 1180, when he was fifteen, may have served as a model for it, but at a deeper level it responds to the human need for the hero to be an outsider, a stranger, which suits his semi-divine status. Arthur was by no means the only leading character of the Matter of Britain to be brought up in exile from his true home and parents. Lancelot, Gawain, Tristram and Mordred were all reared among strangers from infancy, and both Perceval and Galahad grew up in obscurity.

In the middle ages, as earlier, the sons of the aristocracy were sent away from home and brought up in other people's households, evidently in the belief that it is easier to train someone else's boy to be a man than one's own. The corollary in legend is that the hero is reared in the otherworld or in some remote place, cut off from the ordinary human world. Brought up in the ordinary world, he would grow up to be an ordinary man. As it is, he comes to his adult sphere of action from the outside. Instead of quietly inheriting his place in the social structure, or working his way up in it in the normal way, he bursts into it from without and imposes himself upon it.

The coming of Arthur is marked by a supernatural marvel, the test of the sword in the stone. In an early form of this story the sword was Excalibur, the otherworld sword which was known from Celtic tradition to have been made in the Isle of Avalon. It shone with its own light, as brightly as thirty torches. The sword is a masculine symbol, with connotations of virility, action, war and the violent discharge of energy. The anvil, on which the sword is forged, is feminine, passive, obdurate, a symbol of earth and matter, the groundstuff of existence. Drawing the sword from this unyielding sheath is the sign that the hero has emerged into manhood and action in this world from the otherworldly obscurity of his upbringing and moulding. It is a kind of birth and as a result of it Arthur begins to discover who he really is.

The stone in which the anvil is set is probably connected with Celtic traditions linking kingship with sacred stones. Irish kings were crowned on coronation seats which were flat slabs of stone. The most famous one was the Lia Fail, the Stone of Destiny, on which the High Kings of Ireland were crowned at Tara. According to legend, the Lia Fail tested the new king's claim to succeed, for when the rightful king took his place on it, it screamed. It had been discovered when King Conn of the Hundred Battles was walking on the ramparts at Tara and trod on a stone which shrieked many times. The Druids consulted together and told Conn that the shrieks showed how many rightful kings would be descended from him.

The Lia Fail acquired a Christian legend, for it was said to be the stone on which Jacob rested his head at Bethel, when he dreamed of angels going up and down a ladder which stretched from heaven to earth. There was a story that in the sixth century the Lia Fail was sent to Scotland for the coronation of an Irish king there, and was never returned to Tara. It became the Stone of Scone, on which the Scots kings were crowned. Edward I of England carried it off to London, and it is now enclosed in the coronation chair in Westminster Abbey. On the other hand, many people maintained that the Lia Fail never left Ireland, and many that the stone which Edward I seized was not the true Stone of Scone.

After he had been crowned, Arthur and his supporters took up arms against Loth and the rebel lords. Arthur, who had the benefit of Merlin's invaluable advice on strategy and tactics, fought like a lion in battle after battle. Away from the battlefield, he unleashed his energies in other directions. A girl named Lyonors bore him a bastard son, Boarte, who grew up to be a knight of the Round Table. When Arthur went to help King Leodegan of Carmelide, who was beset by enemies, he saw Leodegan's daughter, Guinevere. She was the most beautiful woman in the world and the memory of her stayed in Arthur's mind. Where Carmelide was is not clear, but it was presumably in Cornwall, traditionally Guinevere's home.

The rebels finally gave up the struggle and made their peace with Arthur. Loth sent his wife Morgause on an embassy to Arthur at Caerleon. She came with her four sons – the eldest of whom, Gawain, was now ten years old – and an escort of knights and ladies. She too was a beautiful woman and, not realizing that she was his half-sister, Arthur made love to her. Merlin told him that he would have a son by her, Mordred, who would be born on May Day and who would destroy him and all his knights.

Hoping to rid himself of this threat, in a manner uncomfortably reminiscent of King Herod in the Bible, Arthur later sent for all the children who had been born on May Day. The ship bringing babies from Scotland was wrecked in a storm. Mordred, however, was cast up on shore, where a kindly countryman found him, took him home and brought him up until he was fourteen. It was naturally assumed that he had died in the shipwreck and Loth, seeking revenge, renewed his war against Arthur. When the two sides met in battle, Loth was killed by one of Arthur's allies, King Pellinore of the Isles, who toppled him from his warhorse and struck him a blow that carved through his helm and his head to the eyebrows. The young Gawain swore revenge on Pellinore, and took it ten years later.

Earlier, in Geoffrey of Monmouth, Mordred was Arthur's nephew. The new story, involving Arthur in incest with his half-sister Morgause, goes back at least to the Vulgate *Merlin*. Not much is made of it there and it is easier to believe, because it happens earlier on, when Arthur is still ignorant of his true parentage. More emphasis is put on it in the *Suite du Merlin*, which intended to make Arthur's eventual downfall and death at Mordred's hands the punishment for incest. The fact that Arthur slept with Morgause unwittingly no more allows him to escape the penalty than similar innocence enabled Oedipus in the Greek myth to escape the vengeance of the gods. He is the victim of chance and mischance, of the grim consequences of an accident which sets in motion an inescapable sequence of events. The attempt to snuff out Mordred's life is bound to fail, and there is tragic irony in the fact that Arthur, who has no legitimate son to succeed him, will be brought down by the son he has begotten in incest.

The story creates a curious parallel between Arthur and Mordred. Both are born of an act that outrages morality. Both are taken away from their natural homes and reared in obscurity. It is as if Mordred were Arthur's shadow or double or evil self, who will one day replace Arthur by seizing his throne and his wife, and thereby bring them both to ruin. We are not told that Morgause was ignorant of her relationship to Arthur. On the contrary, in the *Suite* (though not in Malory), it is implied that she knew he was her half-brother. This makes Mordred the child, on his mother's side, of a perverse passion which could well be felt to imbue him with sinister force.

It seems significant that Mordred's birth is so firmly pinned to May Day, which by the old Celtic reckoning was the first day of summer. This may be a piece of seasonal symbolism, looking ahead to the end of the story, with Arthur in his last days linked with winter and the young Mordred, his supplanter, with summer.

Another episode early in the story which looks forward to the end is the finding of Excalibur. Arthur broke the sword which he had drawn from the stone in combat with Pellinore. To replace it, Merlin took him to a lake, where they saw the great sword Excalibur, held up by a hand rising from the water. It was a gift from the Lady of the Lake, a powerful fay who lived in an enchanted palace in the depths of the lake. The scabbard had the magic power of preventing its wearer from bleeding. It would have saved Arthur from ever suffering a mortal wound, if he had not been robbed of it by the treacherous Morgan le Fay.

It was known that at the end of Arthur's life on earth Excalibur would be thrown into a lake and a mysterious hand would rise from the water, catch the sword and brandish it, and draw it down into the depths. To explain this and provide Excalibur with a satisfyingly symmetrical history, the author of the *Suite* arranged for Arthur to break the sword he had drawn from the stone. The hero could then receive his true sword, Excalibur, which would serve him all his life, from the enchanted land at the bottom of the lake to which it would ultimately return. There is also a pleasing irony in the story of the Lady of the Lake giving Arthur the magic sword, for she will soon deprive the king of the magic on which he has relied so far, by taking Merlin away from him. Excalibur's otherworldly origins are clearer here than when it was the sword in the stone and it still comes to Arthur from a feminine source, now the Lady of the Lake instead of the anvil.

Presently, Arthur's barons told him it was time for him to marry. He fixed his choice on Guinevere, the bravest and fairest woman he knew. Merlin tried to warn him that the consequences would be tragic, but only in obscure hints which Arthur failed to understand.

Arthur and Guinevere were married with great splendour at Camelot. An unexpected arrival at the wedding feast was the Lady of the Lake. Merlin fell besottedly in love with her and gave her no peace. She detested Merlin, but concealed the fact and, by playing on his longing for her, coaxed him into teaching her the secrets of his magic. When she left Camelot he went with her. Then she turned his own enchantments against him and sealed him up alive in a tomb in the Perilous Forest. Merlin was never seen again.

In this way, once Arthur had taken a wife, he lost, to a woman, the father figure who had guided his career. Guinevere was to prove an inadequate substitute. Brave and beautiful though she was, she failed to bear Arthur a son and heir, and her passionate infatuation with Lancelot was the cause of Arthur's downfall.

Meanwhile, Arthur had entrusted Excalibur in its magic scabbard to Morgan le Fay, his half-sister. Morgan secretly hated Arthur and waited for an opportunity to destroy him. By enchantment she made a false Excalibur and a false scabbard, which looked exactly like the real ones, and bided her time.

One Monday morning, Arthur went out hunting and pursued a stag deep into the forest. In the evening he came to a river, where he saw a ship, hung with silk down to the waterline. The ship sailed in to the bank at Arthur's feet and he saw that there was no one on it. Unable to resist an adventure, he climbed on board. At once a hundred torches sprang to light and twelve maidens appeared and greeted him prettily. They served the king a sumptuous dinner in a handsomely appointed cabin, after which he went to bed and slept deeply. But the ship was a silken trap. When Arthur woke up, he found himself in a dark dungeon, with twenty other knights. A magic ship which sails by itself appears quite frequently in the romances as a way of transporting a hero to an otherworld or realm of enchantment. This one was sent by Morgan le Fay for her own sinister purposes. Monday, the day of the moon which rules the sea-tides, is traditionally a suitable day for magical operations connected with water and ships.

The castle in which Arthur found himself imprisoned belonged to an evil lord named Damas, who was in dispute with his brother over an estate. The dispute was to be settled by combat between champions representing the two brothers. Arthur was told that he could win his freedom and release the other prisoners by fighting for Damas. This put him in a moral dilemma, for it meant defending a wrongful cause, but reluctantly he agreed. The rival champion was a knight named Accolon, Morgan le Fay's lover. She had given him the true Excalibur and she now sent the false sword to Arthur.

In the combat Arthur fought with courage and skill as always, but his sword was useless against Excalibur and he was repeatedly wounded. The spectators agreed that they had never seen a knight who had lost so much blood fight so well. Bewildered and despairing of his life, Arthur summoned up all the strength he had left and struck Accolon a fierce blow on the helm. It knocked him down, but Arthur's sword snapped in two and he was left without a weapon. Accolon called on him to yield, but Arthur replied that he would rather die with honour than live with shame.

Fortunately for the king, at this critical moment the Lady of the Lake appeared and by magic made Excalibur fly out of Accolon's hand. Arthur seized it gratefully and struck Accolon such a blow that the blood spurted out of his nose, mouth and ears. Accolon surrendered and explained that the whole plot had been the work of Morgan le Fay. Arthur spared his life, but Accolon died of his wounds four days later. Arthur sent his body to Morgan as a present.

Soon afterwards, Morgan contrived to steal the magic scabbard while Arthur was asleep. She hoped to steal Excalibur as well, but the sleeping Arthur held the sword in too tight a grip. Morgan made off with the scabbard, escorted by her knights. Hotly pursued by Arthur, she threw it into a lake and escaped capture by turning herself and her knights and their horses into stone statues. When Arthur saw the statues, he thought that God had turned Morgan to stone in punishment for her crimes. As soon as he had gone, however, she brought herself and her escort back to life, and took refuge in one of her castles.

In this episode Arthur entrusts Morgan with his sword, which is a symbol of his heroic energy and manhood, and in the combat with Accolon he is, in effect, fighting against his own superhuman prowess. The motif of a hero in danger from an enchantress, who desires him, hates him, or both, recurs over and over again in the romances. It is related to the old and widespread theme of the evilness of woman, which is linked with her sexual allure. The enchantress wants either to kill the hero, as in this case, or more often to do away with him metaphorically by keeping him prisoner in her own realm, so preventing him from pursuing his career in the world. She entangles him in a web of mindless sensual pleasure, in which he loses his capacity for action. The encounter with the *femme fatale* is one of the perils which the true hero must experience and survive.

There are several other stories of Arthur falling into the power of an enchantress. According to one (in the Prose Tristan), a sorceress came to Arthur at Cardiff and by promising him an exciting adventure persuaded him to ride with her to her castle in the forest. When they reached the castle she slipped a magic ring on to his finger. The ring made him forget Guinevere and the sorceress seduced him. He might have stayed with her indefinitely, lost to his kingdom and his responsibilities, if the Lady of the Lake had not divined what had happened. She sent one of her maidens, who pulled the ring off Arthur's hand and told him to cut off the witch's head. The sorceress called for her brothers, who overpowered the king and were about to behead him when Sir Tristram came to his rescue in the nick of time.

Arthur is snared here as a result of his knightly inability to reject an opportunity for adventure. In another story (in the Vulgate *Lancelot*) he falls victim to plain lust. While he was fighting the Saxons, he succumbed to the wiles of the Saxon king's sister, the beautiful enchantress Camille. She enticed him to her enchanted tower, where she immured him in a dungeon. By sending word of the king's whereabouts to his anxious court, she succeeded in decoying Lancelot, Gawain and two other knights to the tower and took them captive as well. Imprisonment and separation from his adored Guinevere drove Lancelot mad, and Camille, thinking him virtually destroyed, let him go. The Lady of the Lake restored Lancelot's sanity and gave him a magic ring which broke all spells. Lancelot returned to the enchanted tower, forced his way in with the ring, routed Camille's knights and set Arthur and the others free. Deep in a vault they found a book and a chest. Immediately Camille lost her magic powers and she committed suicide by throwing herself off the tower.

This episode, which shows Arthur in a far from admirable light, was intended to justify Lancelot's love affair with Guinevere. It is typical of a tendency in the romances to weaken Arthur's character. The writers of the romances were more interested in Arthur's knights than in the king himself, for their adventures offered more scope for creativity than the supposedly authentic history of the great king which had been mapped out earlier. As the spotlight of attention focused on the knights, so Arthur tended to fade into the background. Inevitably he shrank in stature and in some stories he became a do-nothing king, idle and poor-spirited.

As the king faded into the middle distance, so did his battles. Writers on the Continent were less enamoured of the tradition of Arthur as the conqueror of Europe than writers in England. When the whole Matter of Britain came to be compiled, a mass of stories about the knights, including the quest of the Grail, had to be fitted in before the final tragedy. As a result, Arthur's military triumphs declined in importance. They were no longer the principal events in the legend. Arthur was now the leading character of the Matter of Britain at the beginning of the story, and again at the end. In between he was relegated to the background, with the foreground occupied by the champions of the Round Table.

The Knights of the Round Table

Arthur enlisted so many knights of renown in his household and the rivalry between them was so keen that he had the Round Table constructed to avoid disputes over who should sit higher in the king's hall. It was said that a skilled carpenter in Cornwall suggested the idea to Arthur and made the table for him. The king could take it with him wherever he went. Since it seated a great many knights, this presumably means that it could be taken to pieces and reassembled. (The famous Round Table now in Winchester Castle appears to date from the 1340s, when King Edward III planned to recreate the Arthurian

Order; he dropped the idea, however, and founded the Order of the Garter instead.)

This is the best-known story of the Round Table's origins, which appears in Wace and Layamon, but there was another story which linked it directly with the Grail. Joseph of Arimathea, the first Keeper of the Grail, was inspired by the Holy Spirit to set up a Grail table in imitation and commemoration of the table of the Last Supper. At the Last Supper the first Mass was celebrated by Jesus himself and the disciples drank the wine, which was the Saviour's blood, from the Grail. Judas Iscariot, however, already had treachery in his heart and he hurried out into the night to betray Jesus. In token of this, one place at the Grail table was always left empty to represent the seat of Judas.

The Grail table in turn was the model for the Round Table, which was made for Uther Pendragon on Merlin's advice. It seated 150 knights. Again, one place at the Round Table was always left empty. This was the Siege Perilous or dangerous seat, and only the supreme hero who was to win the Grail could safely occupy it. In the troubled situation after Uther's death the knights of the Round Table went to serve Guinevere's father, Leodegan of Carmelide, taking the table with them. When Guinevere married Arthur she brought the Round Table to Camelot with her as her dowry. Arthur was delighted and commissioned Merlin to choose the knights to be appointed to it. Merlin did so, predicting the quest of the Grail and telling the knights that the same harmony should reign among them as among the apostles of Jesus. When the knights went to do homage to Arthur, the name of each knight was found written on his seat at the Round Table in letters of gold; the Siege Perilous alone remained empty.

This account of the Round Table, most of which comes from Robert de Boron, displaced the earlier story that it had been made for Arthur himself. The table here belongs as much to Guinevere as to Arthur, and this helps to justify Lancelot's devotion to her. Guinevere is the Round Table's presiding goddess, as it were. Arthur acquires it by marrying her, in something of the same way perhaps as in Celtic tradition he 'married' the land of Britain in her person.

The Round Table is no longer a mere piece of furniture, a neat device to avoid quarrels over precedence. It is an exalted order of knighthood with a Christian mission. Through its link with the table of the Last Supper, it is a symbol of the fellowship of Christ and his disciples, an image on earth of the ideal society of heaven. Its roundness is now interpreted as an emblem of the whole world. 'For in its name it mirrors the roundness of the earth, the concentric spheres of the planets and of the elements in the firmament; and in these heavenly spheres we see the stars and many other things; whence it follows that the Round Table is a true epitome of the universe.'[3]

The universe at this time was believed to be a circular arrangement of spheres, one inside another like the skins of an onion, with the earth at the centre. The Round Table is a microcosm, a miniature image of the whole creation, and so it stands for wholeness, totality, perfection. It is not yet quite perfect, though. There is a gap in the circle. One place at the table is empty and the circle will not be complete until the coming of the Grail hero.

The champions of the Round Table are knights errant, or wandering knights. They ride out from their headquarters at Arthur's court in search of adventure. In *Yvain*, a story by Chrétien de Troyes, one of Arthur's knights is asked to account for himself, and he says: 'I am, as thou seest, a knight seeking what I cannot find; long have I sought without success.' He is asked, what is it that he cannot find? 'Some adventure whereby to test my prowess and my bravery.'[4]

The knight errant is for ever testing himself and measuring himself against odds. His love of adventure is rooted in an insatiable need to stretch his nerve to the limit. By testing and proving himself he gains renown. Chrétien's characters, somewhat comically to the modern eye, are determined to make sure that news of their exploits is carried back to the Round Table for the admiration of the other knights. The hero sets out to win the acclaim of his peers and the love of his lady. In winning them, he helps the unfortunate, rights injustices and makes the world a better place. And yet he often seems to be searching for some objective beyond these immediate aims, for something less transient and harder to gain.

What this objective is, the stories leave us to decide for ourselves. The theme of the quest, the search, is central to the Arthurian romances. The knight errant represents the human impulse not to stay safely at home in tamely familiar surroundings but to strike out into unknown country. He is man as a seeker of risk. The dangers he encounters take many different forms, but the two great perils with which he is threatened are dishonour and death. The true object of his quest is to face and vanquish these dangers. By willingly going out to meet them, he risks his life for his ideal of what he ought to be. It is only in this way that he can achieve his ideal. What he is really searching for, it seems, is his true self, a perfect integrity of character, welded under the hammer-blows of danger. If he succeeds, he becomes a man pre-eminently worthy of honour, whose fame will live on for ever. But there is a nobler prize still to be won. In the quest the hero finds something supremely valuable not only to himself but to others, a life-enhancing treasure which he brings as a gift to his fellow men. His ultimate triumph is a victory over death.

A good example is the story of the Fair Unknown. The earliest version of it that has survived is *Le Bel Inconnu* by Renaud de Beaujeu. The hero is Guinglain, the son of Gawain by a fay, but at the outset he does not know his name or who his father is. He was brought up by his mother alone and she called him Fair Son.

Guinglain went to Arthur's court at Caerleon, rode into the hall and asked the king to grant him whatever he asked. Arthur amiably agreed and, since the young knight had no name and was strikingly handsome, decided to call him the Fair Unknown. Suddenly another stranger arrived, a girl named Hélie, escorted by a dwarf. She asked Arthur to send a knight with her to rescue her mistress, Blonde Esmerée, the Queen of Wales, who had been turned into a dragon by two sorcerers. She could be released from the enchantment only by a kiss. Guinglain volunteered at once and Arthur, bound by his promise, had no choice but to send him. Hélie objected to being given a young, untried champion without even a name to recommend him. She rode away from court in a fury and Guinglain had to hurry to catch her up.

Hélie's attitude changed before long because the Fair Unknown showed himself a brave and resourceful companion. He won them passage at the Perilous Ford by overcoming the redoubtable knight who guarded it; he saved a girl from two giants; he defeated three knights who attacked him. Presently he and Hélie came to the Golden Island, which was reached by a causeway defended by a formidable warrior. This knight hoped to marry the lady of the Golden Island. She did not care for him and would not accept him until he had held the causeway against all comers for seven years. He had now carried out this duty for five years and a row of helmeted heads impaled on spikes testified to his efficiency. Guinglain, undeterred, challenged him, fought him and killed him.

On the island was a palace with crystal walls and seven black towers, and a garden of trees and spices where flowers bloomed and birds sang all the year round. The lady of the island, La Pucelle aux Blanches Mains (the Maiden of the White Hands) was a fay of matchless loveliness, as radiant as the moon gliding out from behind a cloud in the night sky. Unknown to Guinglain, La Pucelle had long been in love with him. She welcomed him to the island and announced her intention of marrying him. Guinglain was powerfully attracted to her, but Hélie reminded him that his duty was to rescue Blonde Esmerée. Next morning, he and Hélie stole away from the island.

They came to the castle of Galigan, where by custom a night's lodging had to be won by jousting with the castellan. Guinglain felled the castellan, who then gave them a kindly welcome. The next day he led them to the Waste City of Senaudon, where the unfortunate Blonde Esmerée was immured, and told Guinglain that when he was welcomed in the city he must reply with a curse.

They reached Senaudon in the evening. It had once been a fine city but it was now in ruins and apparently abandoned. Guinglain rode in alone through the broken gate, past crumbling towers and through eerily deserted streets to a vast palace of marble. At each of its windows stood a minstrel with a lighted candle, playing and singing. The minstrels called a welcome, but Guinglain, as instructed, cursed them. He rode into the hall and halted by a massive table. A knight came from a dark room and attacked him. Guinglain drove him back but was assailed by axes, wielded by no visible hands. Then a huge knight rode at him on a horse which breathed fire. Guinglain stood his ground and killed the knight, whose body turned into a mass of mouldering corruption in front of him.

The minstrels left, slamming the windows shut and taking their candles with them. Guinglain waited nervously in the darkness, keeping his courage up with thoughts of La Pucelle. A glow of light spread through the hall. It came from the jewelled eyes of a horrible fire-breathing serpent, which glided towards him and kissed him on the mouth. A mysterious voice told him that his name was Guinglain and he was the son of Gawain. Overjoyed and exhausted, Guinglain fell asleep on the table.

When he woke up, the hall was full of light and by his side was a beautiful woman, though she was not quite as lovely as La Pucelle. This was Blonde Esmerée, the dragon who had been restored to her human form by the kiss. She told Guinglain that the two enchanters, Mabon and Evrain, had bewitched her and the city to make her marry Mabon, and that the spell had driven the inhabitants of the city away. Mabon was the knight on the fire-breathing horse, killed by Guinglain the night before. Now that she was free, she intended to marry Guinglain.

Guinglain agreed, but in his heart he longed for La Pucelle. Eventually he went back to the Golden Island. There he and the fay consummated their love, and she told him that she had watched over him since boyhood. She had sent Hélie to seek a champion at Arthur's court, knowing that Guinglain would undertake the adventure, and hers was the mysterious voice which had told him his name. His success in the quest had given him the right to her love.

When news arrived that Arthur had announced an important tournament, the fay knew that she could hold Guinglain no longer. That night he went to sleep in her arms, but he woke up in a wood with his horse and his armour beside him. At the tournament he acquitted himself nobly and was reunited with Blonde Esmerée. They went to Senaudon. The people of the city had now returned and Guinglain and Blonde Esmerée were married and crowned king and queen with tumultuous rejoicing.

At the beginning of this story, Guinglain does not know who he is. The quest is a search for his identity and only by braving its perils and hewing resolutely to his duty does he discover his real self. But this is not all. When he frees Blonde Esmerée from the enchantment, he breaks the spell on the city and restores it to life. The sorcerers, Mabon and Evrain, represent the power of death. They kill the city, in effect, by driving the people away and turning it into a ruin. The minstrels with their corpse-candles and their avid, clutching welcome which must not be returned are the walking dead. Mabon himself is dead, and so his body putrefies instantly when Guinglain defeats him. Guinglain brings the Waste City back to life and 'marries' it by wedding Blonde Esmerée, which Mabon had plotted to do. He becomes its king, a king of life instead of a king of death. The people greet him with gratitude and acclaim because he has won a victory over death.

He has also won La Pucelle. She is gained by valour, and this is true of the warrior who defends the causeway as well as Guinglain. Eventually, it is Guinglain's devotion to valour which breaks La Pucelle's hold on him, and he marries the human queen and city, not the fay. It is here that he achieves his full integrity. The fay is a snare. In her way she is another representative of death, and the road to her island is set with severed heads. The hero's true path lies in this world, the human world, not in the other world island with its beautiful palace, its delectable garden and its ravishing mistress.

Many of the story's motifs, including the island, the fay and the *fier baiser* or 'serpent kiss', which breaks the spell, come from Celtic tradition. Senaudon is Snowdon and the Waste City is Caer Seint, the old Roman garrison fort of Segontium at the foot of Snowdon, near Caernarvon. The ruins of Roman architecture made on later generations a profound impression of forlorn grandeur, of a high civilization tragically lost, of life overtaken by death.

The importance attached to Guinglain's real name is characteristic of the Arthurian romances, in which names have a magical mystique. The story-tellers often conceal the names of characters until the last possible moment. The characters themselves frequently keep their real names secret from each other, though to refuse to reveal one's name to an honourable knight is an insult. All this stems from the ancient belief that the real name of a person contains the essence of his being. It is not a mere label attached to him; it *is* him. There is consequently an old fear that magic may be worked against a man through his real name, and a deep, primitive inclination to conceal it.

As the Fair Unknown story demonstrates, the staple occupation of an Arthurian hero is fighting, the activity in which the perils of dishonour and death are most obviously and immediately encountered. The knights of the Round Table fight in war, in tournaments and on quests. They fight to defend the right, to protect the weak, to punish a crime, for the honour of a lady, to avenge an insult or settle a dispute, to test and prove themselves. When one knight errant meets another, his customary greeting is a challenge to joust. There is a description in *Perlesvaus* of Gawain after he has been long away from court on a quest, a figure of weary and battered menace. 'The knight sat on a tall horse, lean and bony. His habergeon was all rusty and his shield pierced in more than a dozen places, and the colour thereon was so fretted away that none might make out the cognizance thereof. And a right thick spear bore he in his hand.'[5]

Fighting is the knight's highroad to the personal integrity he seeks because, like the primitive Celtic heroes from whom he is spiritually descended, it is in violent conflict that he feels most fully himself. Only then is every facet of his being brought into play. Brain, nerve and muscle, courage, skill and experience are deployed with a rapidity of decision and a sense of certainty lacking in the complexities and hesitations of the rest of life. In combat the hero is freed from his misgivings about himself, his sense of his own inadequacy, and he can attain a rapture of self-fulfilment in which he seems superhuman. Like a god or a wild animal, he transcends normal human categories of value. Balin and Balan, fighting for Arthur against Loth, did such shining execution that people could not decide whether they were angels from heaven or demons from hell. When Galahad went into battle against heavy odds, those who saw him thought he was not a human being but a monster. Lancelot in a tournament was like a hungry lion. At another tournament Arthur admiringly compared four of his knights to a mad lion, a ravening leopard and a pair of eager wolves.

The grimness of the hero's apotheosis in combat did not pass unnoticed. Tristram's mother died in giving birth to him. Looking at her baby as she lay dying, she said : 'Ah, my little son, thou hast murdered thy mother. And therefore I suppose thou that art a murderer so young, thou art full likely to be a manly man in thine age.'[6]

The Arthurian heroes are not bovine men, too stupid to feel fear and too stolid to be a prey to self-doubt. On the contrary, they tend to be highly strung, emotional and sensitive. They know what fear is and have to steel themselves against it. They can be knocked off their psychological balance, as well as their horses. When the redoubtable Saracen knight Palomides sees that Tristram will outdo him in a tournament, he bursts into tears. Lancelot and Tristram, parted from the women they love, are so distraught that they go mad. There is no suggestion that such behaviour is unmanly and unfitting. A man who lives by stretching his nerve is likely to be a mass of nerves. A man who only feels fully himself in action feels inadequate out of it. This is as true in the Arthurian romances as in real life.

The Prose Tristan introduced an amusing and subversive new character to the Round Table, Sir Dinadan. Dinadan deplored his fellow knights' determination to seize every opportunity for a fight and though he could give a good account of himself when he had to, he preferred to avoid unnecessary encounters. Cowardice, he remarked, had the great advantage of keeping a man alive. He also could not understand why his brother knights toiled through dangers and hardships to win the favours of haughty, high-nosed ladies when there were plenty of pretty women of less demanding character with whom to enjoy the pleasures of love.

Dinadan had numerous opportunities for stating his heretical views because he was the friend and companion of Tristram, whose thirst for a fight was insatiable. Asked to help Tristram attack no less than thirty opponents at once, Dinadan said that one against two or three was heavy enough odds, but one against fifteen was ridiculous. Tristram dragged him into the mêlée all the same. Later, weary and sore from this adventure, Tristram and Dinadan came to a castle where, by custom, they could obtain hospitality for the night only by defeating the castle's champions. Dinadan said that if that was the custom, he would prefer to spend the night somewhere else, but Tristram insisted on fighting. After vanquishing the champions, the two friends were comfortably settling into the castle when Palomides and Gaheris arrived, demanding a place to sleep. Dinadan would have welcomed them in, but Tristram said that he and Dinadan were honour bound to keep up the custom of the castle by challenging the new arrivals. Dinadan grumbled and mumbled unavailingly. In the joust he was badly bruised by Palomides. He stopped fighting, said that he had never met anyone madder than Tristram, except possibly Lancelot, and rode away in a huff.

Dinadan is subversive because he does not share the knightly ideal of attaining integrity through struggle and conflict. The other characters regard him, with puzzled amusement or irritable incredulity, as a jester and cynic. For them, a knight's honour is inseparable from the duty and pleasure of constant combat. Medieval audiences demanded plenty of fighting in a good story. Modern readers are likely to weary sooner of tournaments and jousts, but some of them still stir the blood, especially when the hero is fighting against odds in a good cause.

Palomides the Saracen went to the Red City to avenge King Hermaunce, who had been treacherously murdered by two wicked brothers, Helius and Helake. Before dying, the king had promised his land to anyone who would avenge him by killing the murderers in one fight. When Palomides arrived, he found the brothers puffed up with pride. They told him they would make him wish he had been christened, and Palomides replied that whenever he died, he would die a better Christian than they were. Then the two brothers rode full pelt against Palomides. The Saracen struck Helake through his shield and armour into his breast and killed him. Helius drove at Palomides with his spear, hurled him off his horse and rode over him, but Palomides seized his opponent's horse by the bridle and pulled down, both horse and man. They continued the fight on foot, lashing at each other with their swords, and Palomides was outmatched. He felt faint and weary, and Helius drove him about the field.

> Then when they of the city saw Sir Palomides in this case they wept and cried and made great dole, and the other party made as great joy.
>
> 'Alas,' said the men of the city, 'that this noble knight should thus be slain for our king's sake.'
>
> And as they were thus weeping and crying, Sir Palomides, which had suffered an hundred strokes, and wonder it was that he stood on his feet, so at the last Sir Palomides looked about him as he might weakly unto the common people how they wept for him, and then he said to himself, 'Ah, fie for shame, Sir Palomides. Why hang you your head so low?' And therewith he bore up his shield and looked Sir Helius in the vizor and smote him a great stroke upon the helm and after that another and another, and then he smote Sir Helius with such a might that he felled him to the earth grovelling. And then he raced off his helm from his head and so smote off his head from the body.
>
> And then were the people of the city the merriest people that might be. So they brought him to his lodging with great solemnity, and there all the people became his men.[7]

Though Palomides had won the lordship of the city, his loyalty to Arthur and his career as a knight errant forbade him to keep it, and he resisted the citizens' persuasions to stay and rule them. The fact that Palomides was an Arab, incidentally, was not held against him at the Round Table, which evidently did not suffer from racial prejudice. His brother knights regretted only that so doughty a warrior had not yet formally joined the Church.

Much of the fighting in which Arthurian heroes engage occurs, ostensibly at least, for the sake of a lady. Knightly attitudes to women in the romances vary enormously, from servile humility at one extreme to sadistic brutality at the other. Generally, a knight fights better if he is in love. Passion lends strength to his arm and he will fasten a love token, given him by his lady, to his shield or his spear point. The token is preferably something that has touched her body – a sleeve from one of her dresses, for example – and a magical power is sensed in its effect on the confidence and prowess of the knight who wears it. The hero's determination to test himself in adventures is sharpened by his need to show himself worthy of his lady's love, but very often what matters to him even more than her love is the fact that she inspires him to deeds of heroism which he values for their own sake, because through them he achieves integrity.

The code of chivalry taught that brutal treatment of women, especially women of rank, was wrong. Lancelot was once told of a knight who ambushed ladies, robbed them, and raped them. 'What?' said Sir Lancelot, 'is he a thief and a knight? And a ravisher of women? He doth shame unto the Order of Knighthood, and contrary to his oath. It is pity that he liveth.' Nor did this evil knight live much longer, for Lancelot soon ran him to earth and dealt him a blow which cut his head in half.[8]

On the other hand, there are examples of a very different attitude. Palomides championed the cause of a girl against a knight who had seized her estates. He met the knight in a joust to decide the issue, defeated him and cut off his head. The girl was delighted and fell in love with her champion. This happened in Surluse, where Prince Galahalt was holding a tournament, and to add spice to the occasion Galahalt announced that any knight who bested Palomides in the tourney should have the girl for himself. Next day, Galahalt himself rode into the field against the Saracen. They drove together so hard that their spears splintered to pieces and then they fought with swords until Galahalt struck Palomides a blow which skidded off his helmet and cut off the head of his horse. Galahalt was deeply ashamed to have done such a thing, even by accident. He apologized profusely and told Palomides he could keep the girl. The contrast between the importance attached to the horse and the girl is revealing.

Another story of this kind turns on the duty of keeping one's word. The beautiful Yseult, with whom Tristram was in love, was the wife of King Mark of Cornwall. An old admirer of hers from Ireland, a lordly knight named Gandin, came to Mark's court and Yseult urged the king to show him every honour, which Mark gladly did. Gandin was a skilful violinist and, after dinner, Mark asked him to play. Gandin said he would play only if he knew in advance what his reward was to be. Piqued by the implied reflection on his generosity, Mark replied that anything he had was at Gandin's command. Gandin played and then demanded Yseult as his reward. Mark was trapped. Unable honourably to break his word, he surrendered the weeping Yseult to Gandin. She was rescued by Tristram, who told Mark to be more careful in future. The story makes it clear that Gandin was guilty of a dishonourable trick, but the implied attitude to Yseult is again revealing and the episode has an unmistakably sado-masochistic appeal.[9]

Many stories of the Round Table centre on the tension between the knight's roles as lover and as fighter. An example is *Erec et Enide* by Chrétien de Troyes. The hero, Erec, was a knight of the Round Table and the son of the King of Estregales (Farther Wales). He married a beautiful girl named Enide and all seemed set fair for them; but Erec became so besotted with his lovely wife that he gave up tournaments and jousting. Instead he mooned over Enide all the time, kissing her and fondling her and staying in bed with her until midday. (An Arthurian knight normally rose with the sun and was likely to have his armour on before breakfast.) People began laughing at him behind his back, and saying that Enide had deprived him of his manhood.

Soon Enide heard the rumours and told Erec what was being whispered about him. Wounded and angry, Erec set out to prove his mettle and test Enide's loyalty by taking her with him on a long journey, in which they met robbers and giants and a lecherous nobleman who wanted Enide for himself. Through all the dangers they encountered Enide remained steadfastly loyal to Erec and he showed himself a champion of more than adequate courage. They were reconciled,and their love was now the richer for being founded not only on sexual passion but also on mutual respect. The lesson is that the troubadour ideal of the knight as the slave of his lady robs him of his integrity and both of them of the possibility of true and lasting love.

Erec and Enide's last adventure on their journey is recounted in the mysterious 'Joy of the Court' episode. They come to the town of Brandigan, the stronghold of King Evrain on an island in a deep and rapid river, where there is an enchanted garden. No stranger has ever penetrated it and come out alive. Despite Enide's distress and the entreaties of King Evrain, Erec is determined to try the adventure. In the beautiful garden, where it is summer all year long and the birds sing sweetly, he sees a row of sharpened stakes, each topped by a severed human head. These are the heads of the champions who went there before him. There is one empty stake, waiting ominously for Erec's head, and on it hangs a horn. No one has ever been able to blow this horn, but whoever succeeds in sounding it will win honour and fame above all other men.

Erec goes further into the garden and sees a beautiful woman sitting on a silver couch in the shade of a sycamore tree. Her lover, a gigantic knight in red armour, appears and challenges Erec. They fight savagely until the red knight falls, exhausted, and yields to Erec. The red knight's name is Mabonagrain. It is he who has killed the previous champions, and he is held captive in the garden by his beautiful mistress until a knight comes and defeats him. He has kept his word to her by fighting his best, but he is only too glad to have been beaten at last, and he will be set free when Erec blows the horn. Erec takes the horn from the stake and sounds it loud and long. Mabonagrain is freed and King Evrain and all the townsfolk are beside themselves with joy – the 'Joy of the Court'.

After this, Erec and Enide returned to Arthur's court, and word came that Erec's father had died. Erec succeeded him as King of Estregales and Arthur gave Erec and Enide a magnificent coronation ceremony at Nantes in Britanny.

The most interesting part of the story is the Joy of the Court episode. Mabonagrain's bondage to his mistress in the garden parallels Erec's earlier besotted preoccupation with Enide. By freeing him, Erec extends to him the liberation from lust's tyranny which he and Enide have already won for themselves. However, this scarcely accounts for the mysterious nature of the garden, the supreme honour which Erec wins in the adventure and the delirious rejoicing with which his success is hailed.

The garden in which it is always summer is a magic Celtic otherworld. It seems likely that the original story lying behind this episode was one of the type in which the hero invades the otherworld to free a prisoner or seize a talisman of regeneration. The talisman in this case may have been the horn and the Joy of the Court (*Joie de la Cort*) may once have been the Joy of the Horn (*Joie del Cor*). The hero's exploit is greeted with supreme honour and rejoicing because he has achieved integrity by risking his life willingly in the garden; and by freeing Mabonagrain from the otherworld he has scored a victory over death.

There are many parallels between *Erec et Enide* and Renaud de Beaujeu's Fair Unknown story, *Le Bel Inconnu*, which was apparently written a few years later. In both there is an island with a magic garden where there is no time and the flowers bloom and the birds sing all the year round. It is approached past rows of severed heads on stakes, showing that it is a realm of death, and is defended by a warrior who is in bondage to a beautiful woman. In *Erec* the knight's name is Mabonagrain and the king of the island is Evrain. In *Le Bel Inconnu* the two sorcerers are Mabon and Evrain. In *Erec* Mabonagrain is bewitched by the mistress of the garden. In *Le Bel Inconnu*, on the other hand, Mabon and Evrain have bewitched the heroine. In both stories, however, the hero breaks a spell to set a prisoner free, in both he wins a battle against death, and in both he is finally crowned as a king. To be crowned king, symbolically, means to be recognized as someone raised above the ordinary human level, someone who has achieved an ideal integrity of character. Kings in Celtic tradition were closely and magically linked with the life and fertility of their lands, and kingship is consequently a particularly suitable prize to be gained by a triumph over death.

In another story, *Yvain*, Chrétien de Troyes returned to the theme of the conflicting demands which action and love made on a knight. Yvain, like Erec, was a knight of the Round Table and a prince, the son of King Urien. In search of adventure he rode into the enchanted forest of Broceliande in Britanny. He came to a clearing where wild bulls were fighting and saw, sitting on a stump with a massive cudgel in his hand, a gigantic and hideously ugly black hunchback, clothed in bulls' hides. The monster was not unfriendly and showed Yvain the path to a magic spring, where the water boiled and bubbled though it was as cold as marble. Overhanging the spring was a tall pine tree. On it hung a golden bowl, and beside the spring was a large stone. Yvain scooped up water from the spring and poured it on the stone. The immediate result was a colossal and violently destructive thunderstorm, with terrifying flashes of lightning, gale-force winds and torrents of rain and hail. When the storm died down, he saw a multitude of birds, covering every branch and twig of the pine tree and singing, each with its own note, in a joyful harmony finer than any music ever heard.

Then there came a knight, riding more swiftly than an eagle and as ferocious as a lion. His name was Esclados the Red and he defended the spring for his wife, the Lady Laudine. He and Yvain charged each other at once and fought a grim, unyielding duel until at last Yvain cut his opponent's head open. Mortally wounded, Esclados fled to his wife's castle nearby, with Yvain close on his heels. Yvain was trapped in the castle but was saved by a girl named Lunete, who gave him a magic ring which made him invisible. The corpse of Esclados was brought into the room where Yvain was concealed and in the killer's presence the dead man's wounds began to bleed afresh, but as Yvain could not be seen he could not be caught. While hiding in the castle, Yvain saw the beautiful golden-haired Lady Laudine and fell in love with her. She was grief-stricken for Esclados but, as Lunete pointed out to her, she urgently needed a new husband of proven mettle to defend her magic spring. Skilfully coaxed by Lunete, she agreed to marry Yvain.

This is the first part of the story. Yvain was originally a real man, Owain son of Urien of Rheged, a North British warrior of the late sixth century, renowned in Welsh poetry as a 'reaper of enemies'. Though he lived after the real Arthur's time, in legend he became one of Arthur's men. There is a shorter and less sophisticated Welsh version of his adventures in the story 'The Lady of the Fountain' (in *The Mabinogion*). Here the monstrous master of beasts who shows the hero the way to the magic spring is even more deformed and sub-human than in *Yvain*. He has only one leg, and only one eye, which is in the middle of his forehead. In both versions he seems to indicate that the hero has entered the realm of wild nature, a region of primitive, mighty and uncontrollable forces which generate a rich diversity of forms, including sports and monsters. The birds whose ravishing harmony combines a multiplicity of notes show that nature is a unity made up of infinite variety. The tree of life and the spring or fountain of life are symbols of immortality in pagan and Judeo-Christian traditions alike, and in medieval descriptions of paradise there is often a huge tree which shelters the souls of the dead in the form of birds. In the story the tree and the spring imply that the hero is close to the hidden sources of life.

Pouring water on a stone to make rain is a piece of imitative magic used by Breton peasants in time of drought down into the nineteenth century. There is apparently an interplay of opposites in the water of the spring which is cold but boiling, the destructiveness of the storm which brings the life-giving rain, and the sweet singing of the birds after the storm. Laudine is the fay or goddess of the spring, and the hero killing her man and taking his place in her bed suggests the theme of an ageing king being replaced by a younger, more vigorous successor to ensure the fertility of the land. Underlying the whole episode there seems to be an affirmation of life victorious over death.

Soon after Yvain and Laudine were married, however, Arthur came to visit them. With him he brought Gawain, the most valiant of his knights and an old friend of Yvain. Gawain urged Yvain not to stay too long with his wife, away from tournaments and manly feats of arms. If he did, Gawain said, he would grow soft and he would forfeit Laudine's respect and love. Yvain decided to go back to Britain with Arthur and Gawain, but he promised faithfully to return to Laudine within a year.

In Britain, Yvain plunged into a hectic round of tournaments and before he knew it more than a year had gone by. A message came from Laudine, denouncing him publicly as a traitor, liar and hypocrite, and saying that she would have nothing more to do with him. Yvain was so ashamed that he went mad. He ran away from court and lived in the forest like a wild beast. He was rescued by a lady who healed him and wanted to marry him, but he refused. Riding moodily through a wood one day, he saw a lion and a dragon fighting. Deciding that the lion was the nobler animal, he came to its aid and hacked the dragon to pieces. The grateful lion promptly adopted him as its master and followed him everywhere like a dog. Its presence at his side induced an attitude of respectful caution in people who encountered him from then on.

Yvain was still desperately ashamed and unhappy. He thought of killing himself, but then he discovered that Lunete, the girl who had saved his life, was about to be burned at the stake for her part in persuading Laudine to marry him. This roused him from his self-pity and, incognito as the Knight with the Lion, he saved Lunete. With the help of the lion, which was a bonny fighter, he also rescued a girl from a giant and set free three hundred maidens who had been imprisoned in the Castle of Evil Adventure. Finally he fought a heroic combat against Gawain to settle a dispute between two sisters who were quarrelling over their inheritance. Gawain was also incognito and, not recognizing each other, the two friends battled on until evening, when both were exhausted. Only then did each discover who the other was, and each then generously insisted that the other had won. Arthur intervened, declared the honours equal and settled the dispute between the sisters. Yvain returned to Laudine's castle. Through Lunete's intercession, Laudine forgave him and they lived happily ever after.

Both Yvain and Erec are publicly humiliated, Erec for abandoning knightly pursuits for love, Yvain for doing the opposite. Each of them has failed to keep a proper balance between the demands of action and love. When shame drives Yvain mad, he loses his human nature and becomes virtually a wild animal. It is a woman he has betrayed and it is a woman who rescues him from his animal condition. He then acquires his faithful lion, which is evidently a descendant of the one that befriended Androcles in the Latin fable. The lion represents the virtue of loyalty and is a sign that Yvain's character is on the mend. All Yvain's subsequent exploits involve him in risking his life to save helpless women from oppression. By turning his back on tournaments and devoting himself to righting wrongs in this way, he achieves true integrity. The proof of this comes when he fights an honourable draw with the great Gawain, the very sun of chivalry. Yvain's success finally cancels out the bad effects of Gawain's advice, which had drawn him away from Laudine in the first place, and Yvain is now worthy to return to her.

The failure of two knights to recognize each other in combat, incidentally, is a constantly recurring device of Arthurian fiction which is not as far fetched as it may seem. A knight in armour with his visor up could not be identified by his face, and his helmet would tend to distort his voice. If he wore plain armour and carried a plain shield, he would not be at all easy to identify.

The mingling of natural and supernatural in *Yvain* is typical of the Arthurian romances. Geoffrey of Monmouth set Arthur and his court in real surroundings but later writers, who were not pseudo-historians, preserved much more of the atmosphere of the Celtic legends. Their heroes live in a country which never was, a country half real and half fairyland. The names of real places are sometimes used, but the geography is weird and a knight will ride from England to Brittany without the formality of crossing the Channel. In Arthur's realm of Logres (from *Lloegr*, the Welsh word for England) the king and his knights move from the natural to the supernatural plane and back again without any sense of strain or incongruity being expressed, either by the story-teller or by the characters themselves. It does not in the slightest surprise a champion of the Round Table to encounter, a day or two's ride from Caerleon or Carlisle, an enchanted castle or a ship that sails by itself or a hideous crone who turns into a beautiful girl. Eventually the romance writers explained that these supernatural phenomena were caused by the mysterious presence of the Grail, concealed in its castle somewhere in Logres.

The supernatural perils which the hero has to meet are tests of his nerve and prowess, just as the natural ones are, but their effect is to make him a man who moves resolutely through a nightmare towards a goal which he cannot see. Erec or Guinglain fighting off robbers know their objective and can calculate the odds against them. Erec entering the enchanted garden or Guinglain riding into the Waste City know only that they must bear themselves well whatever may happen. This powerfully intensifies the central theme of the knight errant as a seeker of danger for its own sake, through which alone he can become his finer self.

The episode of the Castle of Evil Adventure in *Yvain* is an example. Yvain comes to a town in the evening and makes for the castle, looking for hospitality for the night. The people shout rudely at him, telling him that if he lodges there he will regret it, but they will not say how or why: only that he will find out if he is foolish enough to go to the castle. Yvain is naturally angry with them. An elderly lady politely explains that the townsfolk mean him no harm, in fact quite the contrary. She earnestly advises him not to spend the night in the castle, but again does not say why. Yvain answers her in the true style of a knight errant: 'my wayward heart leads me on inside, and I will do what my heart desires'.[10]

At the castle gate a surly porter delivers another cryptic warning. Yvain goes in and comes to a yard enclosed with pointed stakes. In it are three hundred girls, ragged and half starved, toiling over embroidery in the medieval equivalent of a sweatshop. Yvain asks the porter who they are. The porter refuses to tell him and will only say that it is too late for Yvain to escape from the castle now. Yvain asks the girls themselves and discovers that they come from the Isle of Maidens, whose king sends an annual tribute of thirty girls to the castle. Many good knights have been killed trying to rescue them. The custom of the castle is that any champion who stays the night there must fight single-handed against two formidable warriors, the sons of a demon by a human woman. Only when they are killed will the girls be set free. It is not until this point, when it is too late for him to withdraw even if he wanted to, that Yvain discovers the nature of the peril that confronts him. In the morning he fights and kills the demonic brothers, sets the captive maidens free and ends the tribute and the castle's evil custom.

The motif of jousting for hospitality occurs frequently in Arthurian stories. The hero comes to a castle where, to win a night's lodging, he must defeat the lord of the castle or his champion. The knights of the Round Table regard this as a perfectly acceptable rule, which provides them with opportunities to show their mettle, but some castles, like the one in *Yvain*, have sinister customs which it is the hero's task to end.

When Tristram escorted Yseult from Ireland to Cornwall, to marry King Mark, they came to the Weeping Castle, which was apparently somewhere in the Scilly Isles. They hoped to rest there, but were taken prisoner by Brewnor, the lord of the castle. They then discovered that, by custom, when any knight came to the castle with a lady, a beauty contest was staged between the lady and Brewnor's wife. The loser had her head cut off. Similarly, the knight had to fight Brewnor and the loser was beheaded. Tristram was shocked by this custom, which he denounced as shameful and foul, but he felt confident that the lovely Yseult had nothing to fear from any comparisons and he reckoned to give a good account of himself in a fight.

In the morning the two women were shown to the people of the castle. Tristram made Yseult turn round three times and Brewnor did the same with his wife. The people decided that Yseult was even fairer of face and figure than their own lady. Brewnor acknowledged that his wife had lost and told Tristram to cut off her head, as the custom dictated, saying that when he had beaten Tristram in combat he looked forward to taking Yseult as a replacement. Tristram hesitated, but decided that Brewnor's wife deserved to die. He decapitated her with one back-handed sweep of his sword. Then Tristram and Brewnor fought, charging each other like wild boars. Brewnor made the mistake of closing with Tristram and trying to wrestle with him, but Tristram was a big man – even bigger than Lancelot, we are told, though not as strong in the wind. He threw Brewnor to the ground, unlaced his helmet and cut off his head. This put an end to the evil custom, to the great relief of the people of the castle.

Stories like this, which show the hero in his capacity as a righter of wrongs, imply an aristocratic and individualist view of the world. The evil custom is always something that has persisted for years. The people hate and fear it but can do nothing. They can only wait for a hero to arrive and put a stop to it. He does so by accepting the custom, winning the contest and hoisting the villains with their own petard. It is like a game which must be played by the rules. Evil has to be beaten at its own game, and only a hero can do it. People in the mass are helpless.

As in modern thrillers and Westerns, the hero is a man who takes the law into his own hands to right wrongs. He appeals to the instinct in all of us to do the same thing, an instinct which we normally repress, constrained by convention, self-doubt and a civilized awareness of the chaos that might ensue.

Galahad put paid to another evil custom at the Castle of Maidens. When he came to it, seven brothers rode out and attacked him, but Galahad defended himself with such ferocity that they ran away. He rode over the drawbridge into the castle, where an old man handed him the keys and a crowd of girls welcomed him as their deliverer. The wicked brothers had come to the castle ten years before. They were hospitably entertained by the lord but they planned to take his daughter by force, and when he objected they killed him, seized the castle and took prisoner every girl who passed by. Now that the girls were freed, they wanted Galahad to stay at the castle, for fear that as soon as he left the seven brothers would return. Galahad wished to be on his way, so he summoned the knights from the country round about and made them swear never to let the evil custom be revived. The knights were summoned by the blowing of an ivory horn, whose notes echoed through the whole countryside. Soon afterwards news came that the villainous brothers had been killed by Gawain, Gareth and Yvain, and the girls left the castle to return to their homes.

In *The Quest of the Holy Grail* this adventure is related to Christ's descent into hell to free the souls there from the Devil's clutches, and also to liberation from the tyranny of the seven deadly sins, but a Christian gloss has been put on what was originally a pagan story about a hero rescuing prisoners, not from the Christian hell but from a Celtic otherworld. The ivory horn which is blown when Galahad has freed the captives recalls Erec's sounding of the horn in the enchanted garden when he has broken the spell and released Mabonagrain. The girls' vain attempt to keep Galahad in the castle suggests the theme of the hero being tempted to stay in the otherworld.

The Castle or City of Maidens appears quite frequently in the Arthurian romances, and the Castle of Evil Adventure in *Yvain* is a form of it. The maidens may be held there against their will, as in *Yvain*, or they may be seductresses who tempt the hero to abandon his career. In the Galahad story they seem to be a mixture of both. Welsh tradition placed the Castle of Maidens at Gloucester, and Scottish tradition put it at Edinburgh. It probably has its origins in Celtic legends of otherworld islands inhabited by beautiful women. There is a story about Perceval coming to a wide river with a walled town and castle on the far bank. He crossed the river by a bridge and went into the castle, which was completely deserted. He saw a brass gong and struck it repeatedly until the whole building shook to its clangour and a horde of girls suddenly appeared and welcomed him. He was hospitably received by the lady of the castle, who told him that it was the Fortress of Maidens. When Perceval woke up the next morning, the castle had vanished and he was lying in the open.

Castles and walled towns like this are otherworlds or realms of enchantment which are cut off from the world of everyday life by their battlements and frequently by a river or a water barrier of some kind. First and foremost among them is the Castle of the Grail, but there are many others and they preserve the Celtic myth of the otherworld island, reached across water. The Castle of Maidens in the story about Galahad stands beside the River Severn. The Golden Island in the Fair Unknown story is separated from the human world by a stretch of sea, spanned by a causeway. The town of Brandigan, where the enchanted garden is in *Erec et Enide*, is on an island in a deep river. To cross the sea or the river is to leave the firm ground of ordinary human experience and make a passage over water – deep, fluid, swirling, secretive, deceptive – into a mysterious country. It is not necessarily an evil country, but it is uncanny and potentially dangerous, charged with magic force. From a psychological point of view, the water-crossing is a journey into strange, unexplored territory in the mind, which the hero must negotiate in his quest for his true self. He will be tempted to stay there, but it is essential that he resists the temptation and returns to the human world.

In Malory's version of the Fair Unknown story, the Castle Perilous stands by the sea, 'beside the Isle of Avalon'. An ivory horn appears again, hanging on a sycamore tree, but this time it has to be sounded to challenge a fierce knight who is besieging the castle. It announces the hero's arrival on the scene, rather than his success. The corpses of forty knights hanged on trees in full armour indicate what will happen to him if he fails in the encounter.

The hero, Gareth, won his way into the Castle Perilous. There he fell passionately in love with the beautiful Lady Lyonesse, and she with him. That night he went to bed in the hall of the castle and when everything was quiet the lady came and lay down with him. Before they could consummate their love, however, a grim-faced knight came stalking towards them, armed with a battle-axe and 'with many lights about him'. Gareth leapt up and seized his sword. The knight wounded him badly in the thigh, but Gareth cut off the knight's head before fainting from loss of blood. Lady Lyonesse's screams brought her sister, Linet, out of bed. Linet picked up the knight's head and smeared ointment on it and on the neck where it had been severed. Then she put the head back on the neck and the knight picked himself up as good as new.

The same thing happened the next time that Gareth and his lady found an opportunity to make love. On this occasion Gareth cut the knight's head into little pieces and threw them out of the window, but Linet calmly collected all the pieces, stuck them together and mended the knight as before. She had worked the enchantment to prevent her sister from dishonouring herself by sleeping with Gareth before they were married.

The nightmarish atmosphere of the Castle Perilous is typical of the otherworlds which Arthurian heroes invade in search of themselves. The two sisters, the generously passionate Lady Lyonesse and the calculating Linet, with her eerie magical powers, seem to represent aspects of feminity with which the hero has to come to terms. The wound in the thigh may be a metaphor for the effect of the knight's blow, which is to prevent Gareth from making love.

Like a castle, a garden is a secluded place, an enclave separated off from the rush and scurry of normal life, and it too may be an other-world – remote, mysterious, alluring and dangerous. The Queen of Denmark hoped to trap Arthur in her enchanted garden. It had walls of air and inside were beautiful girls and a tree on which luscious red apples grew all the year round. Knights who went into the garden and ate the apples immediately lost all desire for the life of action and stayed happily in the garden with the girls. Any knight who refused to eat an apple was set upon by twenty warriors and three giants. In the end Arthur and Gawain went to the garden and vanquished these champions, so breaking the spell and setting the prisoners free. The beautiful girls went sadly to live in the Queen of Denmark's stronghold nearby, which was ever afterwards called the Castle of Maidens.

Once again, the theme is the release of prisoners from the other-world and the threat to the hero is feminine. The absence of seasons, and so of time, is a common characteristic of Celtic otherworlds and fairylands, where trees bear fruit and flowers bloom all the year round.

The apple is a symbol of female sexuality, red is the colour of desire and the Queen of Denmark's garden is another honeyed trap in which the man of action may be snared.

Lancelot was riding in a forest when he met a maiden who asked him to help Meliot of Logres, another knight of the Round Table. Meliot had been severely wounded and had no hope of recovery unless a champion could be found who was brave enough to go to the Perilous Chapel and bring back a sword and a piece of cloth which he would find there. Lancelot set off at once and presently came to the chapel, with its surrounding graveyard. Leaving his horse tied to the gate, he went into the graveyard, where he saw thirty tall knights, taller than any man he had ever seen, all in black armour and with their swords drawn. The knights grinned at Lancelot and he felt deathly afraid. He took a firm grip on his sword and advanced on them and they scattered out of his way.

Feeling more confident, Lancelot strode into the chapel, which was dimly lit by a single lamp, and saw the dead body of a knight, covered by a silken cloth and with a sword at his side. He cut off a piece of the cloth and the ground beneath him seemed to move a little, bringing back his fears with a rush. Taking the sword and the piece of cloth, he went back outside. The black knights grimly told him to drop the sword, but he took no notice. Leaving the churchyard, he met a beautiful woman. She told him he must leave the sword behind or he would die, but he refused to leave it. The woman told him that if he had, he would never have seen Guinevere again. Then she required him to kiss her. He again refused, which was just as well, for if he had kissed her he would have died. She was a sorceress named Hellawes, who had long desired Lancelot, but she despaired of winning his love because of his devotion to Guinevere. She had planned to trap him at the chapel, kill him and have his body embalmed, so that she could keep it and embrace it. Lancelot's courage and his refusal to grant her even one kiss had foiled her, and she soon afterwards pined away and died. Lancelot took the sword and the cloth back to Meliot and touched his wounds with them, and Meliot was healed.

The chapel in this story is another enclave of the uncanny, where strange beings and forces are encountered. It is a realm of death. Black is the colour of death and mourning, and the grim knights in black armour are evil ghosts. The sorceress who wants to make love to

Lancelot's corpse is in league with death. The hero quells his fears, braves the perils and brings back from the realm of death life-giving talismans, the sword and the cloth, which save the life of his brother knight.

So perverse is Hellawes, the sorceress, that she uses a chapel for her evil purposes. A chapel stands in holy ground. It ought to be, and often is, a sanctuary, a place of refuge from supernatural evil. In *Perlesvaus* Perceval's sister is riding a mule alone in a dark and shadowy forest. At nightfall she comes to a graveyard. In the dark it is a chilling place and it makes her uneasy. Then she sees, outside the graveyard, black knights fighting with spears and swords, the swords red like fire. She is desperately frightened, and so is the mule, but there is an old chapel ahead of her. Hurrying in, she finds the chapel bright with light, and she stays there safely until the morning. It is explained that the warriors she saw fighting were evil spirits, which had taken the form of knights, dead in the forest, whose bodies had not been buried in consecrated ground. In effect, they were the restless dead, but they could not invade the sanctuary of the chapel.

The chapels in both these stories are in a forest and the forest is the scene of many strange adventures in tales of the Round Table. It harbours witches, fays, monsters, enchanted castles and magic springs, like the one in the forest of Broceliande. It is an area of mystery and danger, of things which are not what they seem. In the forest you are far from home, from fireside warmth and security. In the forest you hear rustlings, like whispers and stealthy footsteps. In the forest you are lost. This is part of the lore of childhood now, but in the past dense forests covered much of Europe and faced grown men and women with the menace of the unknown.

The forest is an otherworld, a realm which man has not tamed. It is not always evil, by any means, for it can be green and beautiful, a place of freedom from the shackles of society and convention, a leafy bower for lovemaking, the arena of the hunting of boar or stag. Yet there is usually something uncanny about it. The forest is the territory of wild nature in both its life-giving and its destructive aspects. Since it is unexplored and dangerous, it has a powerful attraction for the knight errant, whose adventures are set there because there is no straight path through the forest and distractions and entanglements are legion.

One strand in both the magnetic attraction and the terror of the forest is the feeling of its antiquity. It was there before man came. It conceals beings and secrets older, wiser and infinitely more powerful than man. The Druids worshipped in woodland sanctuaries, and reverence for sacred trees and groves is embedded in ancient tradition all over Europe. In modern psychological terms the uncanny forest represents the dark depths of the mind, the tangled growths of the unconscious, 'old' and 'wild' in the sense of being primitive, instinctive, unmastered by reason. The hero who penetrates the forest is entering a region in himself. This he must do if he is to discover and achieve his complete and true self, but the territory he enters is not only potentially rewarding but also highly dangerous. The forest is the place where reason snaps. The word 'wood' in Malory means 'mad', and when Lancelot, Yvain and other heroes go mad, they take to the forest and live like wild beasts.

As the realm of wild nature, of the primitive and instinctive, the forest in Arthurian romances is often another home of the female powers, where the seductive enchantress spins her web. In a typical story, a young knight sets out in search of adventure and rides into a forest. By a fountain there he meets a delectable girl, who agrees to be his love if he will do whatever she wants. Bewitched by her beauty, he foolishly accepts the condition. They wander happily together through the woods until they come to a meadow, where the knight feels strangely drowsy. He lies down on the grass and falls asleep. When he wakes up, he finds himself in an enchanted castle from which he can never escape. In other words, by subordinating himself totally to another's will he has ceased to be a man and has become a captive puppet. Or on another level, he has become the helpless prisoner of something in himself, the desire to escape from real life and take refuge in a dream-land of romantic love.

According to another story, Lancelot rode into the Lost Forest, from which no knight had ever returned. He came to a tower, where spellbound knights and ladies danced an endless ring-dance. There was also a chessboard with pieces which moved by themselves and which could only be checkmated by the best knight in the world. Lancelot sat down at the board, played and won. The spell was broken and the knights and ladies were set free.Here again, the hero releases prisoners from the otherworld, where they are caught in a false paradise of futility, symbolized by the ring-dance. The magic chessboard brilliantly suggests the implacability and formidable resources of the powers with which the hero has to grapple. Every move has a counter, every gambit is foreseen.

These strange dream-countries and eerie landscapes of the mind – the uncanny forest, the enchanted castle, the perilous chapel, the enticing garden – are challenges, gauntlets thrown down to the hero who must prove and discover himself by mastering both the natural and the supernatural, or both the conscious and the unconscious. In them he meets not only beautiful enchantresses but also strange beasts, monsters, giants and dragons. These creatures generally represent forces of evil, destruction and death. Perhaps the oddest of them all is the Questing or Yelping Beast, a symbol of incest. It had the body of a leopard, the head of a snake, a lion's hindquarters and the feet of a hare, and from its belly came a noise like a pack of hounds in full cry. Arthur saw it soon after he had made love to his half-sister and begotten Mordred on her.

The story of the Questing Beast concerns a princess, skilled in magic, who made sexual advances to her own brother. He repelled her in horror. She then gave herself to the Devil, who appeared to her as a handsome man, and on his advice she accused her brother of raping her. The brother was condemned to be eaten alive by dogs. The princess, pregnant by the Devil, was delivered of the Questing Beast, which made the noise of hounds because of the loathsome death which she had inflicted on her brother. She herself was forced to reveal the truth and was executed.

Palomides the Saracen hunted the Questing Beast for years and would not allow any other knight the honour of pursuing it. He had become a Christian by conviction, but he had sworn not to be christened and received into the Church until he had killed the beast. According to some accounts, he never caught it, but one story says that he finally cornered it by a lake, where it had stopped to drink. He pierced it with his lance and the creature gave a terrible cry and sank beneath the water. A huge storm burst over the lake as if all the demons of hell had gathered there.

The unfortunate beast, whose body joins together things which ought not to be joined, is specifically a symbol of incestuous desire and more generally of the perverseness and abnormality of evil, engendered by the Devil. Like the Great Beast of Revelation, which it roughly resembles and which also perished in a lake, it is a monster of chaos, anarchy and the ultimate futility of evil. The yelping hounds in its belly drive it remorselessly on a quest which is hopeless – a search for water to quench its insatiable thirst. The Questing Beast is the negative counterpart of the questing knight errant. It is appropriately hunted by the Saracen, who himself joins together things which ought not to be reconciled, because he is a Christian but still technically an infidel.

Dragons, which are also hybrid and evil creatures, are monsters of chaos and death. They lay waste everything for miles, killing by poison and by fire, burning and smashing people and animals, trees and crops, farms and houses. They are impelled by a vast, malevolent lust to destroy (and psychologically can be seen as embodiments of the same impulse in the human mind). The dragons of northern Europe, belching out flames and smoke, seem to be images of death itself, all-devouring horrors whose fiery breath owes something to the blazing flames of funeral pyres on which dead warriors were cremated on battlefields in early times. Killing a dragon is another example of a victory over death.

Tristram fought a dragon when he went to Ireland, a massive brute which was devastating the country. The Irish king was so helpless against it that he had offered his daughter's hand in marriage to anyone of good family who killed it. Tristram mounted a warhorse, took his sword and a sturdy spear, and rode to seek out the dragon in its lair. Spewing flames and smoke, it came at him. He charged it and thrust his spear deep into its throat, but the dragon hit him with a shock that killed his horse, though Tristram managed to scramble away. The dragon, galled by the spear sticking in its throat, rampaged through a wood, bellowing with pain and trampling trees, until it came to a cliff where it turned and stood at bay. Tristram attacked it with his sword, but the dragon slashed at him with its razor-sharp claws, its breath charring his shield and scorching the grass. Tristram dodged the dragon nimbly until the wound told on it and it sank writhing to the ground. The hero plunged his sword into the monster's heart and it let out a last shattering roar and died. Tristram forced its jaws apart and cut out part of the tongue, to take back with him as proof that he had killed it, but the poisonous fumes from the tongue overcame him and he collapsed. He was nursed back to health and claimed the king's daughter, Yseult, for his master, King Mark of Cornwall.

Giants stand for brute force and ignorance, mindless irresponsibility and the destructiveness of greed and lust (in the mind, again, as well as in the outside world). When they are dressed and armed like rustics, as they often are, they seem to represent an aspect of the common people, contrasted with the aristocratic hero. The giant in *Yvain* is a typical specimen of the breed. His name is Harbin of the Mountain and he lusts after the lovely, gently nurtured daughter of the lord of a walled town. Refused her, he has plundered the citizens of everything they possess and ravaged their lands, and he has seized the lord's six sons. Two of them he has killed and he now threatens to slaughter the other four in front of the town gate unless he is given the girl. When he has her, he will hand her over to his servants and scullions for their pleasure. Harbin, in effect, is a personification of the mob.

The lord and the citizens are cowering in the town and wringing their hands when Yvain arrives with his faithful lion and volunteers to fight the giant. The next day Harbin appears. He wears a shaggy bearskin and is armed with a sharpened stake. With him he brings the lord's four sons, in rags and with their hands and feet bound, mounted on spindly, jaded nags. A vile dwarf, puffed up like a toad with self-importance, walks behind and beats the prisoners with a knotted scourge until the blood runs. Yvain spurs out from the town to confront the giant, the lion at his heels. Harbin taunts him, but Yvain says he has no time to bandy words. Charging the giant, he slices from Harbin's cheek a cut of meat big enough to roast. The giant retaliates with a blow of his stake which crumples the hero over his horse's neck. The lion, displeased, bites the giant savagely in the leg. Bawling with pain like a bull, the giant swipes at the lion and misses. Yvain, recovering, runs Harbin through and kills him, the giant striking the ground with a crash like the fall of a great tree.

The hero slaying a dragon or a giant is carrying out his duty as the upholder of right and order. It was a commonplace in the middle ages that the duty of knights was to defend society by keeping the peace and righting wrong. In the stories this is equally the function of the Round Table. 'For by the noble fellowship of the Round Table was King Arthur upborne, and by their noblesse the king and all the realm was ever in quiet and rest.'[11]

Arthur's knights are a band of brothers, in theory, though in practice there are jealousies, rivalries and factions among them, as in any such group in real life. Generally, however, a hero of the Round Table acts alone. There are tales of the exploits of several knights together, but the most impressive stories are about a man facing danger by himself, with only himself to rely on, for only so can he find himself. This is one of the reasons why Arthur fades into the background in the romances. A king usually has a retinue about him and it is not easy to send him on an adventure alone. The greater the hero, the more solitary a figure he tends to be, and lonely courage is one of the marks of the noblest champions of the Round Table.

Lancelot, Gawain and Tristram

'No knight was ever born of man and woman, and no knight ever sat in a saddle, who was the equal of this man.'[12] Lancelot was the supreme hero of the Round Table and for many story-tellers and their audiences quite simply 'the best knight in the world'. Malory's admiration for him is unconcealed, and of all the characters in the *Morte Darthur* Lancelot is the one who arouses the most heartfelt sympathy and admiration.

Malory's Lancelot is an oak tree of a man, as big in physique as in spirit. His sheer size and strength are impressive. On one occasion, deprived of his sword by a treacherous ruse, he tears a branch off a tree and routs his fully armed opponent with that. Unsurpassed in skill, strength and endurance as a fighter, he is rapid in decision and steadfast in action. Kindly, fair minded, thoughtful for others and gentle with the weak, he is intensely nervous and highly strung. There is no meanness in Lancelot and the profound respect in which he is held never goes to his head. He is often described refusing to intervene in a tournament against a knight who has defeated all comers, because he will not try to rob the champion of the honour he has won.

Though brave as a lion, Lancelot knows what it is to be afraid and has to stiffen his resolution to overcome it. At one point he and Arthur encounter Tristram and Palomides, and Palomides knocks Arthur off his horse. Lancelot is scared of tackling two such formidable opponents, but his duty compels him. 'Notwithstanding, whether I live or die, needs must I revenge my lord Arthur, and so I will, whatever befall me.'[13]

Time and again, when Arthur is in danger or trouble, Lancelot comes to his rescue. His tragedy, and the tragedy of the Round Table, is his passion for Guinevere. The conflict between his loyalty to Arthur and his love for Arthur's wife imposes an intolerable strain on him. His affair with the queen robs him of the supreme honour of winning the Grail and in the end, when Arthur can no longer close his eyes to their liaison, pits Lancelot against the leader he loves and brings Arthur's kingdom to ruin.

Lancelot's origins are misty, but he seems to be an example of a god turning in legend into a man, the god in this case being the Irish deity Lugh. The names of Llwch Llauynnaug and Llenlleawg the Irishman, who belong to Arthur's war-band in early Welsh stories, are probably derived from Lugh Lamhfada (Lugh of the Long Arm). The Welsh word *llwch* meant 'lake' and Lancelot from his earliest appearance in the French romances is Lancelot du Lac, 'of the lake'. Llenlleawg the Irishman is described in 'Culhwch and Olwen' as 'the exalted one of Britain', unfortunately without explanation, but perhaps because the god Lugh was linked with the sun. Llenlleawg takes a leading part in carrying off the cauldron of Diwrnach from Ireland, and he is also involved in Arthur's expedition to seize the otherworld cauldron in 'The Spoils of Annwn', so the prototype of Lancelot is already associated with the quest for a magic vessel which is a precursor of the Grail.

Lugh was a mighty warrior, 'Lugh of the fierce blows'. When the Irish gods, the Tuatha De Danann, were threatened by a horde of ferocious monsters from overseas, the youthful Lugh came to Ireland from the otherworld, where he had been brought up, to help them. As he was master of all the arts of warfare and magic, the Tuatha De Danann made him their leader and under his command they defeated the enemy. He was famous for his beauty and, according to one story, he was fostered and trained as a boy by a goddess. Lancelot too was a great warrior, renowned for his good looks, and he was brought up by a demi-goddess in the otherworld.

According to his medieval legend, Lancelot was the son of King Ban of Benoic, a territory somewhere in western France, which Malory thought was either Beaune or Bayonne. Ban himself was descended from King David, so that Lancelot was given an immensely illustrious ancestry which connected him with Christ. When Lancelot was only a year old, Ban was killed in battle and the little boy was snatched away from his mother by a powerful fay, who took him to her home in the otherworld. In one version of this story (in Ulrich von Zatzikhoven's *Lanzelet*), she was a sea fay who lived in a golden castle on a crystal mountain in an island of ten thousand women, where it was always the month of May, where no one grew old, and where fear, unhappiness, jealousy and anger were unknown. In another version (in the Vulgate *Lancelot*), the fay was the Lady of the Lake, and she lived in what appeared to be a lake, though it was actually a magical illusion which she had created to conceal her home from mortal eyes. This version, of course, explains why Lancelot was called 'of the lake'.

The Lady of the Lake brought Lancelot up, but she did not tell him his true name or who his parents were. In this respect his experience was similar to Arthur's. Other heroes of Irish and Welsh legend, including Cuchulain and Peredur, were trained in wisdom and war by women, and this seems to be a reminiscence of an old Celtic custom in real life. From a psychological point of view, the hero brought up without a father is his own man, as it were. He owes his manhood entirely to himself. At the same time, his upbringing among women apparently implies that he is initiated into the mysteries which lie deep beneath the surface appearance of things, as the Lady of the Lake's home lies deep beneath the surface of an illusory lake.

When Lancelot was eighteen, the Lady of the Lake reluctantly decided that it was time for him to return to the human world. She read him a lecture on the qualities and responsibilities of a true knight. Not surprisingly perhaps, since she was a fay, she said nothing about the knight's obligation to defend the Church. Then she took him to Arthur's court, where she herself provided his arms for the knighting ceremony.

The earliest surviving account of Lancelot's love for Guinevere comes from Chrétien de Troyes, in his *Lancelot*. It is a version of the Celtic tale of Guinevere's abduction and rescue, which itself was based on or assimilated to a nature myth, the struggle between winter and summer. Arthur's court was at Camelot on Ascension Day when the queen was carried off by the evil Meleagant, prince of the land of Gorre. Many knights and ladies of Arthur's household were already held captive in Gorre, a land from which there was no return. Behind the figure of Meleagant can be seen Melwas, the abductor of Guinevere in Caradoc of Llancarfan's version of the story. The land of Gorre is really the Ile de Voirre or Isle of Glass, the otherworld, and the theme of rescuing prisoners from the land of death recurs once more.

Ascension Day, which falls in the spring, the day when Christ left the world, is a suitable time for winter to kidnap the source of life.

Lancelot and Guinevere were already in love, though they had not consummated their passion, and Lancelot set off at once to rescue her. In hot pursuit of Meleagant, he accepted, with only a moment's hesitation, a ride in a cart used for the same purpose as a pillory, despite the humiliation and mockery to which he was consequently subjected. Fighting his way through desperate dangers, he came to a place where he found a comb which Guinevere had dropped, with some strands of her golden hair caught in it. Almost swooning, Lancelot reverently disentangled the precious hairs, raising them over and over again to his eyes and lips in an ecstasy of adoration, as though worshipping a holy relic. Then he put them inside his shirt, next to his heart.

Continuing in pursuit, Lancelot came to a graveyard where he saw a marble tomb, sealed with a stone lid so heavy that it would take seven strong men to lift it. A hermit told him that the lid could only be raised by the hero who was to free the captives from Gorre, the land of no return. Lancelot grasped the lid and lifted it without the slightest difficulty. He asked the hermit who the empty tomb was meant for, and was told that it waited for the same hero.

Eventually, Lancelot crossed into the land of Gorre over a bridge which spanned a cold, black, swift-running stream. The bridge was a long, sharp sword, set on edge, and though Lancelot's hands and legs were cut to ribbons and the pain was agonizing, he resolutely crawled over it. (A sword-bridge across the river of death appears frequently in medieval descriptions of the journey to the otherworld.)

Gaining the far bank, the hero came to a strong tower where Guinevere and the other prisoners were held. He was courteously greeted by Baudemagus, the chivalrous King of Gorre, who had prevented his son Meleagant from raping Guinevere. Lancelot immediately challenged Meleagant. Though weak from his wounds and hard pressed by his opponent, he was inspired by the sight of Guinevere watching from the tower, and defeated Meleagant. It was agreed that the queen and the other captives should go free, on condition that Lancelot met Meleagant in single combat at Arthur's court one year later. This duel was to settle the issue finally, and if Lancelot was worsted Guinevere would return to Gorre.

To everyone's astonishment, Guinevere treated her rescuer coldly.

She was determined to punish him for having hesitated before accepting for her sake the humiliation of riding in the shameful cart. But she soon repented of her cruelty and fell ill with grief. Thinking she was dead, Lancelot tried to hang himself, but was prevented just in time. The lovers arranged to meet secretly by night at Guinevere's window, which was guarded by thick iron bars. Their passion was too strong for them. Lancelot wrenched the iron bars apart by main force and knelt to the queen as to a saint. That night they took their joyous pleasure of each other for the first time.

Lancelot proved his submissive devotion yet again at a tournament when, at Guinevere's whim, he played the coward. After this, Guinevere and the other prisoners went back to Camelot, but Lancelot was waylaid by Meleagant and shut up in a windowless tower, where he was fed sparsely on bread and water. He was released by a woman he had earlier helped, just in time to return to Arthur's court and meet Meleagant in combat on the appointed day. Meleagant was no match for Lancelot in his fury. Lancelot hacked his right arm off and felled him. Meleagant was too proud to beg for mercy and Lancelot cut off his head, to the relief and rejoicing of Arthur and the whole court.

The story is another variation on the theme of the hero who wins a battle against death and whose triumph is greeted with great rejoicing. At the cost of appalling pain Lancelot enters Gorre, the land of no return, or land of death, and liberates the prisoners held there. He subsequently defeats and kills Meleagant, who represents winter as an aspect of death, and so puts paid to the threat that the source of life, represented by Guinevere, will be taken away for ever. In effect, he frees the world from a sentence of death and guarantees the continuation of life.

Chrétien de Troyes does not seem to have felt comfortable with his *Lancelot*. He left it to someone else to finish and he says at the beginning, defensively it seems, that both the story itself and his treatment of it were imposed on him by his patroness, the Countess Marie of Champagne. The reduction of the hero to the status of the petulant Guinevere's lap-dog makes disagreeable reading, though there is psychological truth in it. The story may have been written with a covert irony, intended to disparage the troubadour ideal of the lover as the slave of his lady. Lancelot's adoration of Guinevere is described in terms of religious devotion. The Welsh tradition, apparently, treated Guinevere as a goddess and, ironically or not, in *Lancelot* she is the goddess of the troubadour 'religion of love'.

The ecstatic adoration which Lancelot feels for the queen is another of the hero's escape routes from the prison of his everyday self. In love, as in the joy of battle, he wrenches apart the bars of his cage and soars into an empyrean of the spirit in which he is more alive, more complete, more fully himself.

On the other hand, the story turns Arthur into a weakling. When Meleagant comes to Camelot and defies the king, saying that he has no intention of returning the prisoners he has taken, Arthur replies meekly that he must perforce endure what he has no power to remedy. The Arthur of yore would not have tolerated such an insult for a moment. And it is now Lancelot who goes to Guinevere's rescue, not Arthur, as in the earlier versions of the tale.

It may have been the Countess Marie herself who first thought of making Lancelot and Guinevere fall in love. If so, she planted the seed of one of the world's finest love stories, but on the whole later writers did not care for Chrétien's treatment of it. Bits and pieces of it crop up in later romances, but a different story of Guinevere and Lancelot became the accepted one. When Lancelot first came to Arhur's court to be knighted, the king omitted to gird on his sword. It was the queen who presented him with the sword, a detail which looks forward to their future relationship. Lancelot fell hopelessly in love with the beautiful queen and scarcely dared to look at her, and she found something touching in his shyness.

Lancelot rode out to prove himself in adventures. He came to the fell keep of Dolorous Garde (Castle Sorrowful), built on a rock near the river Humber, where many prisoners were held captive by enchantment. The hero fought his way into the castle, so breaking the spell and freeing the prisoners. In the graveyard he found a tomb covered by a metal slab which could be lifted only by the hero who had conquered the castle. He raised the slab and read the inscription: 'Here will lie Lancelot of the Lake, the son of King Ban of Benoic.' Now, for the first time, he discovered his name and who he really was. The castle was renamed Joyous Garde and many years later Lancelot was duly buried there. Local tradition in the north of England has long identified Joyous Garde as Bamburgh Castle, a massive Norman stronghold on the coast of Northumberland.

Galahalt, the lord of the Faraway Isles (probably the Scilly Isles), arrived at Arthur's court and became a close friend of Lancelot, whom he warmly admired. He discovered Lancelot's secret passion for the queen and told her, as was true, that Lancelot was wasting away for love of her, unable to eat or drink. Galahalt arranged the meeting at which Guinevere and Lancelot exchanged their first kiss. The lovers did not sleep together until the night when Arthur betrayed Guinevere in the bed of the enchantress Camille. Although their adultery is thus given an excuse, Guinevere sadly recognizes that their sin will prevent Lancelot from achieving the Grail.

Lancelot knows that his relationship with the queen is disloyal and sinful, but he will not give her up. In *Perlesvaus* he tells a reproving hermit that it is the sweetest and fairest sin he ever committed and he cannot bring himself to repent of it. If he did, he would destroy himself, for it is the queen's feeling for him that makes him what he is. All his success and renown he owes to the inspiring power of her love. When the hermit tells him that as long as the liaison continues he is a traitor to Arthur and an enemy to God, Lancelot replies: 'So dearly do I love her that I wish not even that any will should come to me to renounce her love, and God is so sweet and so full of right merciful mildness, as good men bear witness, that He will have pity upon us, for never no treason have I done toward her, nor she toward me.'[14]

Besides the burden of his conflicting loyalties to Arthur and Guinevere, Lancelot had also to endure the machinations of Morgan le Fay. Morgan hated Arthur and his queen, did her best to destroy the Round Table and had designs of her own on Lancelot, who was powerfully attractive to women. Once, when Lancelot was away from court in search of adventure, he went to sleep under an apple tree: a dangerous place to choose, because the apple is the fruit of the otherworld and of female enchantments. Morgan came riding by with two other enchantresses. They saw Lancelot asleep under the tree and they all wanted him as a lover. By magic they spirited him away to Morgan's castle, where they told him that he must choose one of them as his mistress or die. Lancelot said he would prefer to die and keep his honour intact. Fortunately for him, a girl in the castle helped him to escape.

There are numerous stories of Morgan's unsuccessful attempts to open Arthur's eyes to his wife's intrigue with Lancelot. Trapping Lancelot, she put him to sleep with a drug and took from his finger a ring which Guinevere had given him and which he had promised always to wear. She sent the ring to Arthur's court with a message that Lancelot was dying and deeply regretted his affair with the queen. Guinevere, racked with anxiety for Lancelot, swore that her fondness for him was innocent, and Arthur believed her. Balked, the enchantress let Lancelot go, but later she captured him again and while he was her prisoner for many weary months he painted pictures of himself and Guinevere on the walls of his room in Morgan's castle. Morgan afterwards showed these pictures to Arthur and though they did not quite convince him that Guinevere and Lancelot were guilty, they made him profoundly suspicious.

Meanwhile, however, Lancelot went to the town of Corbenic, where the Grail was kept in the castle by the lord of the town, King Pelles. The people greeted him joyfully and asked him to help an unfortunate girl who was being boiled in a tub of scalding water. She had been condemned to this fate by Morgan le Fay, who was jealous of her beauty, and she could be freed only when the best knight in the world took her by the hand. Gawain had recently tried to release her, but he had failed. The citizens took Lancelot to a tower and a room as hot as a stove, where the poor girl scalded in her tub 'as naked as a needle', as Malory says. Lancelot took her by the hand and released her.[15]

King Pelles, the lord of Corbenic, welcomed Lancelot to the castle and invited him to a sumptuous banquet, at which a beautiful girl came in bearing the Grail, a vessel of gold. She was Elaine, the virgin daughter of Pelles. The king wanted Lancelot to lie with Elaine and beget on her a son, who would achieve the quest of the Grail. He said nothing to Lancelot, but he hatched a plot. Lancelot was told that Guinevere was at a nearby castle. When he hurried there to see her, an enchantress named Brisen gave him a magic drug in a cup of wine which made him think that Elaine was Guinevere. Lancelot went to bed with her and it was not until the morning that he realized that a trick had been played on him. Furiously angry, he drew his sword and threatened to kill Elaine, but she knelt down naked and begged for mercy: 'for I have in my womb begotten of thee that shall be the most noblest knight of the world.' Lancelot forgave her and kissed her, 'for she was a fair lady and thereto lusty and young, and wise as any was that time living.'

When her time came Elaine was delivered of a son, who was christened Galahad. He grew up to surpass even his father in prowess, and indeed to win the Grail. The parallel between his conception and that of Arthur at Tintagel is obvious.

Elaine loved Lancelot, and never married. Later, when Arthur held a great feast at Camelot, she went to court. She and Guinevere greeted each other politely, but nothing more, for the queen knew what had happened and, though Lancelot had explained the circumstances to her, she was fiercely jealous. She told Lancelot to come to her that night, suspecting that otherwise he would go to Elaine. However, Brisen the enchantress contrived to lead Lancelot to Elaine's bed instead of Guinevere's, and in the darkness he again made love to her thinking that she was Guinevere. When the queen found out, she was beside herself with rage. She told Lancelot that he was a traitor and she never wanted to set eyes on him again. Lancelot was so appalled that he went mad. He jumped out of the window into a thorn-bush, which scratched him badly, and ran away.

Guinevere soon repented. She besought Lancelot's brother, Ector de Maris, and his cousins, Bors and Lionel, to find him. They searched for him in vain, and so did Gawain, Yvain, Perceval and other knights, sent to look for him by Arthur.

Lancelot 'suffered and endured many sharp showers, that ever ran wild wood from place to place, and lived by fruit and such as he might get and drank water two years'.[16] Eventually he came to Corbenic. No one realized who this madman was. The dogs chased him and the young men of the town pelted him. The people of the castle gave him straw to lie on by the gate and threw him scraps of meat from a safe distance. After a time, Elaine saw him and recognized him. Brisen put him into an enchanted sleep and he was carried into the castle to the room where the Grail was kept. The sacred vessel restored him to sanity.

Lancelot was desperately ashamed of his misfortunes, and his only thought was to hide himself away where no one could find him. He and Elaine went to the Joyous Isle, where it was summer all year long, but Lancelot was deeply unhappy and often wept for his loss of Guinevere and Arthur. In the end, Perceval and Ector de Maris, Lancelot's brother, came to the Joyous Isle and Perceval discovered that Lancelot was there. 'Then Sir Perceval sent for Sir Ector de Maris, and when Sir Lancelot had a sight of him he ran unto him and took him in his arms; and then Sir Ector kneeled down, and either wept upon other, that all men had pity to behold them.'[17]

Ector persuaded Lancelot to return to Camelot, where the court greeted him with joy. When Guinevere heard the tale of his adventures, she wept. Arthur, still unsuspecting, assumed that it was love of Elaine that had driven him mad, and so did most of the court, but Lancelot's brother and his cousins knew for whose sake he had gone out of his mind.

Lancelot's experiences with Morgan and Elaine form a counterpoint to his involvement with Guinevere. The theme of a fay or enchantress falling in love with a knight and trying to keep him her prisoner in the otherworld occurs frequently in the Matter of Britain. Elaine is not described as a fay, but she comes from the otherworldly Grail castle and Lancelot takes refuge with her from the human world in the enchanted Joyous Isle, where there is no time.

The fay or enchantress is drawn to the knight by his heroism, which has a magnetic attraction for her, and yet if she has her way she will end his heroic career. She wants to keep the hero as he is, preserved like a fly in amber, and she lures him to her enchanted realm where there is no time and so no change. But the hero needs change and development, because he is in search of his true self. The alluring fay represents the conflict between action and love. This conflict is not resolved by making the hero a man who has no sexual nature at all, as some Arthurian writers did with Perceval and Galahad. The true hero is a whole man, not a eunuch. The true hero is Lancelot, who remained passionately devoted to one woman all his life, and the result was tragedy.

Guinevere does not keep Lancelot from action. On the contrary, her love inspires him to heroism. His relationship with her is the supreme Arthurian example of the conflict between loyalty to one's love and loyalty to others who have a right to it. Because Lancelot is the ideal knight, he is the ideal lover, but because he is the ideal lover he is trapped in a situation in which he falls short of the loyalty of the ideal knight.

Lancelot fails in the quest of the Grail. He is outstripped by his own son, Galahad, and humiliated. He returns to Camelot a chastened man, but in Malory, before the shadows finally close round Lancelot, there is an episode which demonstrates his quality. A knight from Hungary named Urry was brought to Arthur's court. He had been wounded in a tournament and a sorceress had put a spell on him, that his wounds could be healed only by the best knight in the world. Lancelot was away from court and Arthur and many of the other knights touched the suffering Urry's wounds, but to no avail. When they had given up in despair, Lancelot arrived. Arthur told him he must try to heal Urry. 'Jesu defend me,' said Sir Lancelot, 'while so many noble kings and knights have failed, that I should presume upon me to achieve all that you, my lords, might not achieve.' Arthur insisted, however, and Urry himself begged Lancelot to make the attempt. Then Lancelot prayed silently to God and knelt down beside Urry and touched his wounds, and the wounds were healed. 'Then King Arthur and all the kings and knights kneeled down and gave thankings and loving unto God and unto His Blessed Mother. And ever Sir Lancelot wept, as he had been a child that had been beaten.'[18]

Gawain is Arthur's nephew, his sister's son, and Welsh tradition celebrates his fearlessness, his unfailing courtesy to guests and strangers, and his 'golden tongue'. His reputation for persuasive speech and polished tact persisted in the early romances, where he is contrasted with Kay, who is a thorough bully, abrasive, sarcastic and quarrelsome. Gawain is Gwalchmei (possibly 'Hawk of May') in Welsh stories and Gauvain or Gavain in French. The modern form of his name is Gavin. He may have been a real man originally, the real Arthur's nephew and lieutenant, but if so he was apparently connected at an early stage with the Irish hero Cuchulain, who in his own cycle of legends was nephew, sister's son, of the king of Ulster, Conchobar. The relationship between a king and his sister's son is peculiarly close and significant in the Arthurian stories and this appears to be a legacy from early Celtic kinship systems. Tristram is sister's son to King Mark and Perceval to the Grail King. Mordred's relationship to Arthur is a twisted variant of it. If a king has no son by his wife, it seems, his sister's son is his heir.

Cuchulain's father was the god Lugh. Gawain's father was Loth of Lothian, and Loth may be a corruption of Lugh. Lugh was connected with the sun and his feast day fell at the height of summer, at the beginning of August. It is to this link, apparently, that Gawain owed a primitive peculiarity which survived in the romances and in Malory. His strength grew steadily during the morning, as the sun's does, reached its peak at noon, when he was at his most dangerous, and then waned again during the afternoon.

Chrétien de Troyes made Gawain the sun of chivalry. 'He who was lord of the knights, and who was renowned above them all, ought surely to be called the sun. I refer, of course, to my lord Gawain, for chivalry is enhanced by him, just as when the morning sun sheds its rays abroad and lights all places where it shines.'[19]

Like other Arthurian heroes, Gawain was brought up by strangers. According to one version of the story, he was the bastard son of Arthur's sister and Loth. They concealed the birth, put the baby in a barrel with a letter explaining who he was, and floated him out to sea. A fisherman found him and took him to Rome, where he was adopted by the Pope and educated until he was old enough to return home. The story resembles the one about Gawain's half-brother, Mordred.

The earliest surviving tale which deals at any length with Gawain's adventures is Chrétien de Troye's *Conte du Graal*, on which Wolfram von Eschenbach based his *Parzival*. The story tells how Gawain came to a garden where he saw a beautiful woman, sitting under a tree and admiring herself in a mirror. Her name was Orgueilleuse. 'Of her the adventure says that she was the very bait of love, to eyes a sweetness without pain and a bowstring to stretch your heart.'[20] Despite, or perhaps because of, the fact that she treated him with haughty and withering contempt, Gawain was captivated by her and determined to win her.

They rode along together and eventually came to a river. On the far bank was a magnificent castle with five hundred windows, in each of which stood a woman or a girl in brightly coloured clothes. Orgueilleuse went off on business of her own at this point and Gawain was rowed across the river by a ferryman, who told him that he had come to a savage land full of marvels. The castle had been built by a queen, who lived there with her daughter and granddaughter. It was called the Castle of Wonders and was guarded by magic. Five hundred automatic bows fired of their own accord at anyone who attacked it. Its inhabitants were fatherless girls, elderly widows who had been robbed of their lands, and young men who had not yet been knighted. They all longed for a hero to come and break the spell, find husbands for the girls, right the old ladies' wrongs and knight the young men.

Though warned that he was going to his death, Gawain strode into the castle, which was lavishly furnished and decorated with gold and jewels, ivory and ebony. He came to a room in which stood a superb four-poster bed, richly carved and hung. When he sat down on the bed, it shrieked aloud, arrows and crossbow bolts mysteriously flew through the air at him, and a monstrous lion sprang upon him and dragged him to his knees. Gashed by the arrows and defending himself desperately, Gawain succeeded in beheading the lion and cutting off its paws. He had passed the test and broken the spell on the castle.

The Castle of Wonders is another form of the Castle of Maidens, separated from the human world by a river. The queen and her daughter and granddaughter may represent the stages of female life, from maiden to mother to crone. The prisoners in the castle are young men, girls and old women, living in luxurious security but cut off from real life until a grown man comes to rescue them from a state of incompleteness and unfulfilment. It cannot be any man who succeeds in this task, but only a hero who is faultlessly brave, faithful and true. The adventure of the Perilous Bed crops up in several of the romances as a test of the hero's courage and worth. It appears to be another way of representing the conflict between action and love. It is the bed of sensual pleasure and restful ease, which threaten to distract the hero from his career of action. Arrows and darts are the weapons of Eros, the god of love. The lion suggests the overwhelming force of the hero's own animal passions.

Gawain is greeted with great honour by the inhabitants of the castle and it is now revealed that the queen is Ygraine, Arthur's mother, and her daughter is Gawain's own mother, whom he has not seen for twenty years. Gawain is told that he is now the lord of the castle and he can never leave it again. In other words, an attempt is made to keep him in the enchanted world of femininity which he has invaded and mastered, the world of his own mother and grandmother. Being a true hero, Gawain will have none of it. He rides away, back to the world of action, and after many other adventures he wins the love of the beautiful Orgueilleuse.

Just as Kay, who was originally a heroic figure, turned into a rude, barbarous foil to the polished Gawain, so Gawain's character changes for the worse in later stories. His interest in women is excessive and when it is a question of getting one into bed he is totally unscrupulous. In one episode he carries off a girl against her will and says that no one can stop him because he is Arthur's nephew. In another he promises to help Sir Pelleas, who is hopelessly in love with the lady Ettarde.

He then cynically tells Ettarde that he has killed Pelleas and proceeds to seduce her.

When Gawain came to Corbenic, he was begged to rescue the unfortunate lady who was boiling in a tub. The feat could be achieved only by the best knight in the world, and Gawain failed. He was then greeted by King Pelles, the lord of Corbenic, in his hall. While they were talking, a white dove flew into the room. It carried a golden censer in its beak and the sweetest scent filled the hall that any man could imagine. The tables were set for a meal and into the hall came the most beautiful girl Gawain had seen in all his life. She held a chalice, the richest vessel that human eyes could ever see, and as she passed through the hall the most delectable food in the world appeared on the tables.

The girl was Elaine, the Grail Bearer, and unfortunately Gawain was far more interested in the girl than in the Grail. He followed her with his eyes until she left the hall. Then he looked down at the table and discovered to his shame that he alone had been given nothing to eat. In the end, the townsfolk put him in a filthy cart, drawn by a spindly nag, and drove him out of Corbenic, pelting him with dung, mud and old shoes.

Besides turning into a womanizer, Gawain acquired an evil reputation for merciless ferocity. Vengefulness became a ruling trait in his character and it is typical of the new attitude to him that he was made responsible for the murder of Perceval's father and brother. According to the Prose Tristan and the *Suite du Merlin*, Perceval's father was King Pellinore, who killed Gawain's father, Loth, in battle. Gawain, who was then a boy of eleven, swore vengeance and nursed a bitter hatred of Pellinore in his heart. When they were grown up, Gawain and his brother Gaheris treacherously murdered Pellinore. Later they set a trap for Lamorak, Perceval's elder brother, who had meanwhile become their mother's lover. Gaheris burst in on the lovers in bed and cut off his mother's head, but he let Lamorak go because he was not armed and could not defend himself. Lamorak fled from court, Arthur was furious and Gawain was not pleased with his brother's nice discrimination between the lady and the knight. Gawain and Gaheris, with their brothers Agravaine and Mordred, soon afterwards set on Lamorak. He defended himself vigorously but odds of four against one were too heavy for him. Mordred finished him by stabbing him in the back.

One reason for this change in Gawain's character was that his earlier position as the leading knight of the Round Table was usurped by Lancelot. Lancelot was a great lover as well as a heroic fighter, and Gawain's relationships with women were treated in contrast. Another reason was the evil repute of Arthur's sister, Morgause, which spread to her children by Loth, Gawain and his brothers. Their relationship with Mordred also did them no good and they acquired a grim fame for treachery and murder. According to Geoffrey of Monmouth, Loth was lord of Lothian, the south-east of Scotland, and a tradition grew up that his headquarters were at Traprain Law, a huge prehistoric hill fort on the edge of the Lammermuir Hills. But Loth was also the nephew of the King of Norway and later writers shifted the whole family northwards to the Orkneys, which were ruled by Norse earls in the middle ages. Gawain's association with the savage north and the Vikings may have had something to do with his deterioration in the French romances.

In England, however, especially in the north, Gawain remained an admired and favourite character and he is the hero of a splendid Middle English poem, *Sir Gawain and the Green Knight.* The story begins at Camelot on New Year's Day, when there rode into Arthur's hall a gigantic warrior who was bright green. His skin was green, his hair was green, his clothes were green, even his towering horse was green. In one hand he held a holly branch and in the other an immense battle-axe as sharp as a razor. He said he had come to play a Yuletide game. Any champion who dared could strike him one blow with the axe, on condition that a year later the champion submit to a return blow from the green knight. When no one hurried to take up this offer, the stranger taunted the knights as cowards until Gawain accepted the challenge. The green knight calmly lifted up his long hair to expose his neck and Gawain with one mighty sweep of the axe cut his head clean off his shoulders. It rolled on the floor among the courtiers, who nervously pushed it away with their feet. Not in the least daunted, the green knight picked up his head by the hair and turned it to face Gawain. The eyelids opened and the mouth spoke, telling Gawain to meet him for the return blow a year later at the Green Chapel. With that the green knight rode away.

The year passed all too quickly and when the time came for Gawain's departure Arthur's court was plunged in anxiety and grief. Gawain, however, said cheerfully that a man must face his fate, whatever it might be. He rode north in search of the Green Chapel on his famous warhorse Gringolet (a Welsh otherworld horse originally, white with red ears). He reached an unknown country in the icy grip of winter, where he fought his way past dragons, wolves and trolls. Riding through a wild forest on Christmas Eve, he came to a noble castle. He was welcomed by the lord, Sir Bercilak, his lovely wife and a hideously ugly old woman who was her companion. The old woman, though Gawain did not know it, was the enchantress Morgan le Fay.

Bercilak told Gawain that the Green Chapel was not far away, and invited him to stay at the castle until New Year's Day. Bercilak went out hunting each morning, while Gawain stayed indoors, resting from his journey, and they light-heartedly agreed that Bercilak would make Gawain a present of the game he caught in the hunt, and Gawain would give Bercilak anything he acquired in the castle. Each day Gawain stayed late in bed and Bercilak's beautiful wife came to him and tried to seduce him. Though sorely tempted, Gawain resisted her, but he did accept from her a magic girdle of green silk which she said would protect him from all harm. Hoping that it might save him at the Green Chapel, Gawain did not give it to Bercilak.

When New Year's Day dawned, Gawain took a sad farewell of everyone at the castle. A servant, showing him the way to the Green Chapel, urged him to run away and save himself from certain death, but Gawain said there could be no excuse for such cowardice. He rode on alone to the Chapel, which proved to be not a building but a hollow mound in a deep valley beside a waterfall. There was no one there, but presently Gawain heard the ominous sound of an axe being whetted. Then the green knight appeared and Gawain bowed his head for the blow. The green knight raised his axe and Gawain could not help flinching. The green knight rebuked him and they tried again. This time Gawain stood still as a rock and the green knight brought the axe down and hit him a glancing blow which just nicked his skin.

The green knight explained that he was Sir Bercilak, the lord of the castle, transformed by Morgan's enchantments. Morgan had devised the whole adventure in the hope of discrediting the Round Table. Gawain's life had been spared because he had honourably refused to make love to Bercilak's wife, but his neck had been grazed because he had concealed the gift of the girdle. Bercilak told Gawain that he had proved himself the most faultless knight who ever lived.

He pressed Gawain to go back to the castle for a celebratory feast, but Gawain insisted on returning to Camelot. He took the magic girdle with him.

The story ends with everyone happy except Gawain, who feels he has disgraced himself by breaking his word to Bercilak in the hope of saving his life. He resolves to wear the girdle always as a badge of shame. At Camelot, however, he is greeted with proud acclaim and every knight agrees to wear a green baldric as a mark of honour to Gawain, who has brought fresh renown to the Round Table.

The poem celebrates the virtues of Christian chivalry. The hero is courageously content to let the outcome of the adventure rest on God's will. Only when he keeps the magic girdle does he waver in reliance on God and for this entirely understandable fault he is lightly punished, though he himself takes it deeply to heart. Through his ordeal, he gains a more perfect integrity.

For all its Christian tone, though, the poem achieves its main effects with pagan motifs. The beheading game itself had appeared in several earlier stories and came originally from Celtic legend. The oldest surviving version of it is in the Irish story 'Bricriu's Feast', where the hero is Cuchulain and the challenger is a giant. Cuchulain beheads the giant. When it comes to the return blow, the giant raises the axe to the roof of the hall, but he brings it down gently, blunt side downwards, and Cuchulain for his bravery is recognized as the champion of champions. The Irish word used for the challenger is *bachlach* (churl), which may be the origin of the green knight's name, Bercilak.

No precedent is known for the green knight's greenness. The colour was often associated with the fairies. The description of the Green Chapel in the poem is not completely clear, but it seems to be a prehistoric barrow, and these ancient burial mounds were connected in folk belief with the otherworld and the fairy race. There could perhaps be a touch of erotic symbolism in the description of the hollow mound in a deep valley by a waterfall, suggesting female sexuality and the enchantments of the otherworld. The green knight's link with winter, his branch of evergreen holly, his castle in the mysterious forest with its beautiful sensuous chatelaine, all suggest that in the shadows behind him stands a Celtic lord of the otherworld, an immortal master of death and renewed life.

One theory is that the story is based on the motif of the annual combat between winter and summer, but it does not fit easily into this framework. What more likely lies behind it is the theme of the hero who goes to the otherworld, braves its perils and wins a victory against death. Refusing to stay in the otherworld, as Gawain declined to return to Bercilak's castle, he goes back to his own country with a talisman of regeneration and is greeted with great honour and rejoicing. In this case the talisman is the magic girdle of invulnerability, and the regeneration is the fresh sense of pride and confidence with which Gawain's achievement inspired Arthur's court.

Arthur's court, it was said, had a magnetic attraction for bold, fighting men, who flocked to the great king's service. In a sense this was perfectly true, for heroes whose stories were originally independent of Arthur were drawn into his orbit and were made knights of the Round Table. One of them was Tristan, or Tristram as he is usually called in English. There are several parallels between Tristan and Lancelot. Each of them is an orphan, each of them falls in love with the wife of the king who commands his allegiance and affection, and each of them is torn apart by the consequent conflict of loyalties.

The love story of Tristan and Yseult was first told a thousand years ago or more. Tristan was the nephew of King Mark of Cornwall, who was a contemporary of King Arthur and held sway at Tintagel. Tristan's father was a stranger from another country, who was killed in battle before Tristan was born. His mother, Mark's sister, died within a few hours of giving birth to him. Brought up in his father's country by a loyal servant, he grew up to be a redoubtable warrior, a skilful harper and a master of the arts of hunting and falconry. He went to Cornwall and became a favourite of his uncle, King Mark.

A huge Irish champion named Morholt came to Cornwall to demand a tribute of young men and girls as slaves for his master, the King of Ireland. Mark and the Cornish barons were terrified of Morholt, but Tristan fought him and killed him. Morholt's corpse was sent back to Ireland as a sardonic 'tribute' from Cornwall. Embedded in the skull was found a splinter of Tristan's sword. The splinter was kept by Morholt's niece, Princess Yseult the Fair, who swore to avenge him.

Later, Mark sent Tristan to Ireland to find him a wife. The hero killed the ferocious dragon which was laying waste the country and Yseult gratefully tended his wounds. She had learned her skill in healing from her mother, who was an enchantress. While Tristan was

Tristram's Family Tree

According to Malory

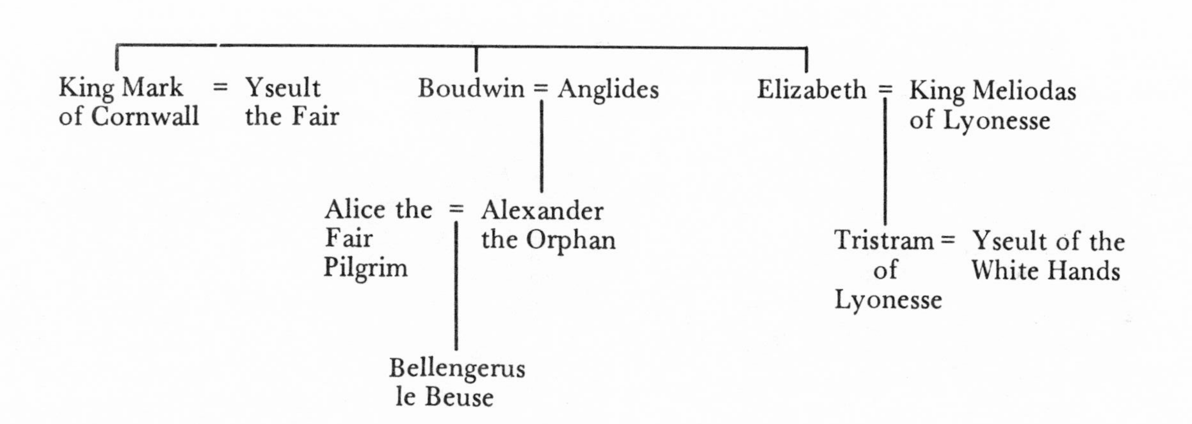

in the bath, Yseult saw the notch in his sword, matched it with the splinter she had kept and realized who he was. She was tempted to kill him with his own sword while he wallowed helplessly in the bath like a stranded fish, but her womanly reluctance to kill proved even stronger than her hatred. When Tristan recovered, it was agreed that he would escort Yseult to Cornwall, where she was to marry Mark. They set off for Cornwall in a ship, on far from friendly terms with each other, but by accident they drank a love potion, made by Yseult's mother and intended for Yseult and Mark on their wedding night. Immediately they fell passionately and irresistibly in love.

In Cornwall, Yseult was married to Mark. To conceal the loss of her virginity, she persuaded her faithful companion Brangane to take her place in bed on the wedding night and in the dark Mark was deceived. Afraid that Brangane might one day reveal the secret, Yseult plotted to murder her, but the plan miscarried. Though Yseult and Tristan pulled the wool over Mark's eyes with considerable skill, the unfortunate king became increasingly suspicious and set traps for them. At one point he forced Yseult to swear her innocence on red-hot iron. She arranged for Tristan to disguise himself as a pilgrim, and when she came to the place of the ordeal she stumbled and the 'pilgrim' caught her. She then swore that no man but Mark and the pilgrim had ever held her in their arms and, since this was true, the iron did not burn her.

Finally Mark became convinced that he was being cuckolded by his dearly loved wife and his dearly loved nephew. According to one version of the story (in Eilhart von Oberge and Beroul), he resolved to burn them both alive, but Tristan escaped on the way to the stake. Mark then handed Yseult over to a band of lepers, to serve their lusts and live with them in degradation as an outcast. Tristan rescued her from the lepers and the lovers fled to the forest. According to a different version (by Gottfried von Strassburg), Mark could not bring himself to kill the lovers and he banished them to the forest. After a time, hunting in the forest one day, Mark found the two of them asleep with Tristan's naked sword between them. They had thoughtfully placed it there before going to sleep, in case of discovery. Mark was deceived and, thinking they must be innocent after all, took Yseult back.

Tristan went into exile in Brittany, where he married the Breton king's granddaughter, Yseult of the White Hands. He married her because she had the same name as Yseult the Fair and because he suspected that his own Yseult had forgotten him and was enjoying herself in Mark's bed. He took no real interest in her and he could not consummate the marriage. Yseult of the White Hands, who loved him, grew fiercely jealous of her namesake.

In the end Tristan was badly wounded in battle by a poisoned spear which, significantly, struck him in the loins. He sent a ship to Cornwall for Yseult the Fair, asking her to come and heal him. It was arranged that if the ship returned with her on board it would hoist white sails, and if not, black sails. Yseult came willingly and the ship drew in towards land under white sails, but the jealous Yseult of the White Hands, who was keeping watch, went in and told Tristan that its sails were black. At this news Tristan gave up his hold on life. When Yseult the Fair reached his side, he was dead, and she died of grief beside him. Their bodies were taken back to Cornwall and buried in separate graves at Tintagel. Two trees grew from the graves and the branches intertwined and joined above them.

The love potion is the key to the story. It is a symbol of love as an overwhelming supernatural force, which seizes on human beings and sweeps them away with it, against their inclinations and despite their good intentions and their obligations to others. Against it they struggle as ineffectually as a lamb in the talons of an eagle. It carries them to heights of rapture which few will ever know. In its grip they break every tie of loyalty and trust, flout every rule of society, lie and cheat and connive, and bring sorrow and ruin on themselves and on people to whom they owe affection. Mark and Brangane and Yseult of the White Hands are as much the innocent victims of the tragedy as Tristan and Yseult themselves. This is the madness of love and those who are not caught in it can only look on with horror and envious longing.

In Gottfried von Strassburg's *Tristan*, on which Wagner based his opera, the lovers' mutual ecstasy is described in terms of religious experience. It is compared to the rapture of the mystic in the embrace of God, in which the soul escapes from the body to lose itself in the bliss of union with the divine. But Tristan and Yseult are human beings, trapped in the body and the world, and their love cannot find lasting and total fulfilment in earthly life. The magic potion is a poison. The elixir of love is the draught of death. 'It is in God's hands!' said Tristan. 'Whether it be life or death it has poisoned me most sweetly! I have no idea what the other will be like, but this death suits me well! If my adorable Isolde were to go on being the death of me in this fashion I would woo death everlasting.'[21] The final part of Gottfried's poem is lost and we do not know how he would have finished the story, but Wagner's opera ends with Isolde singing the *Liebestod*, 'the love-death': in death the two souls are at last made fully one, one with each other and with the eternal life of the universe.

The original Tristan is believed to have been a real man named Drust son of Talorc, who was a king of the Picts in Scotland in about 780. According to a story which may have been founded on fact, he went to the Hebrides, where he discovered that the daughter of the local king was about to be handed over to three pirates or giants from the sea, who demanded her as tribute. Drust dealt with the robbers and saved the princess. This legend belongs to the same general type as the Greek tale of Perseus rescuing Andromeda from a sea-monster. It apparently lies behind the Morholt episode and details from it, including the recognition of the hero in the bath, survived in the medieval story of Tristan.

The legend of Drust is thought to have migrated to Wales, then to Cornwall and Brittany, and then to England and France. On the way, in Wales, the central love triangle of the king, his wife and his nephew was added to it, apparently drawn from the Irish legend of Diarmaid and Grainne, in which the heroine is compelled by a magic spell to fall in love with her husband's nephew. This is presumably the source of the love potion and details from this story also survived, including the episode of the sleeping lovers separated by the sword. In Wales the leading characters became King March (Horse), his wife Essyllt and his nephew Drystan son of Tallwch. It is thought that the dragon-slaying episode, which recalls Perseus again, may have been added when the story reached Brittany. The tale of Yseult of the White Hands, with its theme of the hero loved by two women of the same name, is possibly a mingling of Breton and Arab traditions, and other oriental influences on details of the legend have been identified. The ship's black and white sails may be Breton as well, though this motif also occurs in the Greek legend of Theseus.

If this speculative reconstruction of the Tristan story's growth is right, it is a fascinating example of a powerful legend being gradually put together by successive generations of story-tellers. One problem is that the real King Mark is believed to have ruled in South Wales, probably in Glamorgan. If so, why was the whole story moved to Cornwall? An inscription of the mid-sixth century on a memorial stone near Fowey in Cornwall reads: 'Here lies Drustanus son of Cunomorus.' This Drustanus is not likely to have been the original Tristan, but by the ninth century in Brittany Cunomorus was identified as King Mark. It may be that the hill fort of Castle Dore near Fowey was the stronghold of the real chieftain, a contemporary of the real Arthur, who appears as Mark in the legend. The fort dominates the southern end of the old trade route across Cornwall from Ireland to the Continent and whoever controlled it would certainly have had dealings with Ireland. Though the central elements of the legend were put together in Wales, it may have been set in Cornwall and attached to Cunomorus, identified as Mark, because the memorial stone was erroneously assumed to indicate the grave of the real Tristan.

However this may be, the complete story was known in England and France by about 1150 and several twelfth-century authors wrote versions of it. According to most of them, Tristan's father came from Loenois or Lohnois, meaning Lothian, south-eastern Scotland. In English, however, Loenois became Lyonesse, a legendary country to the west of Cornwall, between Land's End and the Scilly Isles, which like Atlantis had sunk beneath the sea. The legend of Lyonesse was probably inspired by distant memories of areas of land which the sea had actually engulfed. The floor of Mount's Bay in Cornwall was inhabited long ago, before the sea swallowed it. In the fourth century AD, apparently, there was only one Isle of Scilly. Now there are several and at low tide the remains of walls can be seen in the sand-flats between them.

When Tristan was brought into the Arthurian cycle the tragic love story was heavily watered down. The old story turned on the conflict between Tristan's loyalty to Mark and his passion for Mark's wife. In the thirteenth-century Prose Tristan, however, which became the standard version, the events are seen from the viewpoint of Arthur's court. Mark is a determined enemy of the Round Table and all good knights, a black-hearted villain, cowardly and treacherous. Tristan owes Mark no loyalty. On the contrary, in opposing him and trying to rescue Yseult from her wicked husband, he is upholding the standards of chivalry. With one leg of the drama sawed through in this way the whole tragedy collapses and it almost disappears from sight behind the adventures allotted to Tristan in his new role as a knight errant.

One reason for the change was the need to avoid having two similar love stories – of Tristan and Yseult, and Lancelot and Guinevere – both attached to the Round Table. This is suggested by a scene in which Lancelot and Tristan meet. The two great champions do not recognize each other. They fight a heroic joust, with the honours so even that in mutual admiration each presents the other with his sword. They then reveal their identities and Lancelot asks Tristan what he thinks about love. Tristan, who has evidently given the subject no previous thought, says that love is a nuisance and a source of weakness. For Lancelot it is the most precious and strengthening thing in the world.

In the Arthurian version of the story, Tristan and Yseult start to fall in love before they drink the love potion, so that the theme of overwhelming passion striking like a bolt from the blue is undermined. When Tristan first meets Yseult in Ireland, he discovers that Palomides the Saracen is in love with her. Out of sheer bravado he decides to cut Palomides out, and Yseult is favourably impressed. In Cornwall, meanwhile, Tristan has a mistress, the wife of one of Mark's barons. Mark wants the lady for himself and when she rejects him in favour of Tristan, he conceives a deadly enmity for his nephew. This is long before there is any question of Tristan cuckolding him with Yseult.

The story continues, roughly on the lines of the older legend, until Tristan and Yseult run away from Cornwall to England together and find refuge at Lancelot's castle of Joyous Garde. Then Tristan sets out with the other knights of the Round Table on the quest of the Grail. Mark takes the opportunity to make a lightning raid into England, seize Yseult and drag her back with him to Cornwall. Tristan forsakes the Grail quest and goes to join Yseult at Mark's court. He is singing and playing the harp to her one day when Mark bursts into the room and stabs Tristan in the back with a poisoned spear. All Yseult's healing skill cannot save Tristan and, embracing her in his dying convulsion, he strangles her.

There were various accounts of Mark's fate. In one, Tristan killed him before dying. In another, Mark was killed by his great-nephew Bellengerus, in revenge for Tristan. In a third, Mark was shut up in prison and force-fed until he died of over-eating. According to other stories, Mark survived the death of Arthur. He then invaded England and harried the whole country. He wrecked the city of Camelot and smashed the Round Table to pieces. He was finally killed, in one version, by Bors, Lancelot's cousin.

The Arthurian story of Tristan eclipsed the older, more primitive and more powerful legend, which did not re-emerge from the shadows until the nineteenth century. The genuinely tragic character in it is Palomides the Saracen, who becomes the most sympathetic figure in a new love triangle. His unrequited love for Yseult conflicts with his friendship for Tristan. Yseult likes and pities Palomides, but cannot love him. Tristan resents his devotion to Yseult, but is fond of him. Palomides is driven to distraction by his passion for Yseult and his jealousy of Tristan, not only as his rival in love but as his superior in knightly prowess. With part of himself he hates Tristan and yet he cannot help liking and admiring him. There is no way out for Palomides and in the end he is butchered by Gawain, who takes cruel advantage of him when he is already weak from wounds. He dies forgiving Gawain, commending himself to Christ's mercy and regretting only that he had hoped for more time to be of service to God and the world.

Merlin and Morgan le Fay

There was never a real Merlin in the sense that there probably was a real Arthur, but the legend of the great enchanter conceals a Welsh bard and seer of the sixth century named Myrddin, to whom various poems and prophecies were attributed. Although it was said that the town of Carmarthen in Wales was named after him, he seems to have been mainly associated with the north-west of England. Tradition linked him with the battle of Arfderydd, which was 'fought for a lark's nest' in 573. Arfderydd is the modern Arthuret, in Cumbria, north of Carlisle, and the lark's nest was presumably Carlaverock ('the fort of the lark'), commanding the northern bank of the Solway Firth.

At this battle, apparently, Rhydderch Hael defeated and killed Gwenddolau, who was Myrddin's lord. Rhydderch Hael (Rhydderch the Generous) was a king of the British dynasty of Strathclyde, which ruled south-western Scotland from its stronghold at Dumbarton. Gwenddolau seems to have been a Cumbrian chieftain. According to the story, Myrddin felt responsible for the deaths of Gwenddolau and many others in the battle and was so horrified by the slaughter that he went raving mad. He fled to the Caledonian Forest in the Scots Lowlands, where he lived for years as a 'wild man of the woods', and in his madness acquired the gift of foreseeing the future.

In the Welsh poem 'Afallenau' (Apple Trees), attributed to Myrddin, he bewails his miserable existence in the forest. He sees an apple tree beside a river bank. At its foot, when he was sane, he used to sport with 'a fair wanton maiden, one slender and queenly'. Now he is alone, an outcast, starving, pinched with cold, mad and grieving for his lord and his countrymen. 'Death has taken everyone, why does it not call me?'[22]

There are similar Scottish and Irish legends, related to the Welsh one. It is possible that Myrddin was a real man, the bard of Gwenddolau in Cumbria, and that the legend about him had some foundation in fact. He was associated with Carmarthen either because he came from there or through a mistaken attempt to derive the name Carmarthen (Sea Fort) from Caer Myrddin (Myrddin's Fort). An alternative possibility is that the name Myrddin was a fiction to begin with, derived by false etymology from Carmarthen in the same way that a man called Lleon was invented to explain Caerleon: though this supplies no explanation of why the wild-man legend should have been attached to him. In either case, he was not originally the magician who guided the real Arthur. If he was alive in 573, he could not have been more than a boy when Arthur died.

It seems to have been Geoffrey of Monmouth who transformed Myrddin the seer into Merlin the magician, turning the name into Latin as Merlinus instead of Merdinus, perhaps because the latter would have been uncomfortably close to the French word *merde*. Geoffrey introduced Merlin into his *History of the Kings of Britain* with striking effect at the point when the Saxons turned against Vortigern, who had brought them to Britain as mercenaries. Vortigern and his court fled to Wales, where his magicians advised him to build a strong tower as a safe retreat. He chose a site for the tower in the mountains of Snowdonia, but his workmen's efforts to build it were frustrated because the foundations kept sinking into the ground. Vortigern's magicians told him that the only remedy was to find a boy who had no father, sacrifice him and sprinkle the foundations with his blood.

Men were sent to scour the country for a boy with no father, and one was found at Carmarthen. His name was Merlin and he was the grandson of the King of Dyfed. No mortal man had sired him: his father was an incubus, a demon which took the form of a handsome stranger to make love to the king's daughter. After giving birth to the boy she had become a nun.

When the youthful Merlin was brought before Vortigern he treated the royal magicians with blistering contempt and told the king that his tower would not stand because there was a pool in the ground beneath it, a fact which his magicians had been too incompetent to discern. At the bottom of the pool, Merlin said, were two hollow stones, and inside the stones were dragons. Vortigern's workmen dug down and found the pool. From it emerged two fire-breathing dragons, one red and one white, which began to fight.

Merlin explained to the astonished Vortigern that the portent meant that the British (the red dragon) would be hard pressed by the Saxons (the white dragon) until the coming of the Boar of Cornwall (Arthur), who would trample the Saxons beneath his feet. Merlin went on to deliver a long string of prophecies in obscure symbolic language and Vortigern was vastly impressed.

Merlin next comes into the *History* as an adviser to Aurelius and Uther Pendragon. He brings Stonehenge from Ireland and erects it on Salisbury Plain. He uses his magic arts to stage-manage the conception of Arthur at Tintagel and after this he disappears from the story.

The tale about Vortigern and the tower had been told three hundred years before by Nennius, in whose version the boy's name is Ambrosius. Ambrosius Aurelianus was the leader of British resistance to the Saxons and the story may have preserved an authentic tradition of hostility between Ambrosius and Vortigern. Ambrosius is Emrys in Welsh and the legend was set at Dinas Emrys, a hill fort near Beddgelert in Snowdonia, where archaeologists in the 1950s discovered the site of the pool from which the dragons emerged in the story.

Transferring this legend to Merlin backdated him to the period of Vortigern, so that he could be presented as the master-mind who foresaw the coming of Arthur and engineered the hero's entry on to the stage. It also sprang Merlin on the readers with an excitingly romantic shock as the mysterious 'boy with no father', who boldly confounded Vortigern's wizards and saved his own life with a display of superior clairvoyant and prophetic powers. With its contest of rival magicians and its threat of human sacrifice, the story has a primitive and pagan atmosphere. The belief in spirits which fathered children on human women was a pagan Celtic superstition before it became a Christian one. Though Merlin was largely a fiction, he stood for something which was probably common in the real Arthur's time, the state of being half Christian and half pagan, of adopting the new religion without surrendering the old.

Geoffrey of Monmouth's *History* began the process which carried Merlin to fame as the Grey Eminence of the Matter of Britain. Making him the offspring of an unnatural coupling between a pagan spirit and a Christian princess gave him a conception legend parallel to Arthur's own and supplied an explanation of his dominating personality, his superhuman gifts and his unearthly wisdom. Merlin was the product of a union of opposites, of immortal and mortal, pagan and Christian, evil and holy. An eerie atmosphere clung to him ever afterwards in the annals of the Round Table.

The legend of Merlin and Stonehenge is interesting because, although the monument was not brought to England from Ireland, the bluestones which formed the smaller and older circle at Stonehenge did come from the west and they probably were transported most of the way by sea and river, as the story says. (The larger stones, the sarsens, were dragged on sledges from near Marlborough.) As they were in place at Stonehenge by about 1600 BC, the legend in the *History* is based on an accurate tradition some three thousand years old. The bluestones came from the Prescelly Hills in Pembrokeshire and substantial Irish immigration into south-west Wales in the early centuries AD could account for the tale that the stones were brought from 'Ireland'. In the *History* Merlin moves Stonehenge to Salisbury Plain as much by engineering skill as by wizardry, but there is only a thin line between superior technology and magic, and in popular belief his success was an astonishing feat of enchantment. In Wace he moves the stones by casting a spell on them.

Merlin's link with the Giants' Dance again associated him with pagan traditions. Stonehenge was the most impressive pre-Christian monument in Britain, the bluestones came originally from a pagan sacred hill, and the *History* makes Merlin himself say that the stones were connected with 'secret religious rites' or 'mysteries' and that they had magical healing power (a belief which survived locally for centuries). Evidently Merlin understood the religious secrets of Stonehenge. In the later legends, although Merlin is a Christian, he is something more. He represents an older understanding of man and nature, a profound wisdom from the pagan past, not in opposition to Christianity but in anticipation of it. It may be that memories of the Druids went to his making. The Druids did not build Stonehenge, any more than Merlin did, but they were famed for their magical and prophetic gifts. They were priest-magicians, probably with a shamanistic strain in their ancestry, the masters and custodians of an old tradition of arcane knowledge about the world and its workings.

Geoffrey's *History* established Merlin as a prophet of the first rank and the cryptic predictions attributed to him were studied, and added to, for centuries. There is nothing whatever in the *History* about Merlin's being a wild man of the woods. It looks as if Geoffrey at this time knew of the Welsh Myrddin only as a seer, whom he could use to build up a picture of Arthur as a figure of destiny. Apparently it was not until later that he came across the tradition of Myrddin the wild man, who is the central character of his *Life of Merlin*.

According to the *Life*, Merlin went mad with grief over the carnage of a battle and ran away to the Caledonian Forest. His sister's kindly attempts to coax him back into the normal world failed because contact with civilized society was too much for his precarious sanity. In the summer and autumn, when the apples hung on the boughs, he wandered contentedly in the forest glades. In winter he lived in a house which his sister built for him in the forest, with seventy doors and seventy windows. There he studied the stars and delivered prophecies, which were dutifully copied down by seventy secretaries. Merlin is described here as a sage who is close to the divine, and the number seventy may have been borrowed from the Bible as a number of the servants of God: the seventy elders in the Old Testament and the seventy missionaries sent out by Christ in the New.

Geoffrey had now pulled two Merlins out of his hat, though he maintained that the *Life* recorded the later adventures of the Merlin of the *History*. Gerald of Wales and other Welsh writers could not swallow this, and decided that there must have been two different men of the same name: Merlin the enchanter and Merlin the wild man. The authors of Arthurian romances concentrated on the magician and ignored the story of the wild man, but some of the latter's exploits were credited to the former. An example is the tale of Merlin's prediction that a certain boy would die a triple death. He was proved right when the boy fell from a high rock, was caught in the branches of a tree, from which he hung upside down with his head under water, and drowned. Another example is the motif of Merlin's three sardonic laughs. He laughed when he saw a leaf caught in a queen's hair and realized that, unknown to her trusting husband, she had been lying with a lover. He laughed when he saw a ragged beggar, because he clairvoyantly recognized that the poor man was unsuspectingly sitting above a cache of buried treasure. He laughed when he saw a young man buying new shoes guaranteed to last for seven years, because he knew that the young man would die the next day. The second and third of these semi-proverbial occasions for laughter at the grim irony of life also occur in Jewish folklore, where it is the demon Asmodeus who laughs.

The figure of the wild man of the woods has a long history in European legend and folklore. He is an outcast from society who is at home in the uncanny forest. Solitary, hairy, naked or clothed in tatters, he lives on fruit and berries, roots and grubs. He is mad, but there is an old and widespread belief that the mad are closer to the gods than the sane. The wild man, for all his poverty and suffering, is free of the chains and shackles which society fastens on the human spirit. Living in harmony with nature and the animals, he is in touch with forces more fundamental than civilized man allows himself to experience: in modern psychological terms, with the deep forces of the unconscious mind. Like the wild forest itself, he is an ambivalent being, dangerous but benevolent, sinister but holy.

This tradition seems to have influenced Merlin's character in the romances. It reinforced the ambiguity of his nature, half human and half demonic. Like the wild man, he is essentially solitary and a free spirit, answerable to no one, unimpressed by rank and title, careless of convention. He comes into the story always from outside, suddenly appearing out of the blue to give advice or perform some piece of magic, and then vanishing again. Wherever his home is, it is not at Camelot, but somewhere beyond the boundary of the everyday human world. There is apparently a limit to the amount of human company he can tolerate, and in the end he is cut off from human society altogether. He seems ageless and it is not certain that he has ever died.

Later writers were not content to let this fascinating character disappear from the scene when Arthur was born, as in the *History*. They added incidents to strengthen his role as the agent of unseen forces of destiny which moulded the fortunes of Arthur and the Round Table. The contrast between his demonic father and his virtuous mother was sharpened in Robert de Boron's *Merlin*, which says that the magician's conception was planned by the Devil and the lords of hell. They plotted to ruin mankind by bringing into the world a false prophet, half demon and half man, an evil counterpart of Christ. One of the fiends succeeded in lying with a pure virgin in her sleep, but because she was a pious Christian the infernal plan was thwarted. She realized what had happened and consulted her confessor. He made the sign of the cross over her, gave her holy water to drink and imposed on her a vow of lifelong chastity. In other words, the resources of Christian magic were mustered against the machinations of evil. The result was that when Merlin was born, he was gifted with superhuman powers, including the ability to see into the past and the future, but he did not inherit his father's evil will and hatred of mankind. His exceptional hairiness, his cruel sense of humour and his mischievous fondness for practical jokes were legacies from his demonic sire, but though he was enigmatic, uncanny and dangerous to cross he was genuinely devoted to Arthur and his knights.

Merlin was a prodigy. He could speak as soon as he was born and while he was still a baby he dictated the history of the Grail to his mother's confessor, Blaise, who had now become his tutor and secretary. Blaise may be the famous Welsh story-teller Bleddri in a new disguise, but the significant thing is that it is Merlin who transmits the knowledge of the holiest of Christian relics to a Logres still sunk in pagan darkness. By using the ability to discern the past which he owes to his demonic father, and thus employing potentially evil powers for good, he acts as a bridge between the old dispensation and the new, between paganism and Christianity.

An even more important addition to the story, on the same lines, is that Merlin persuades Uther Pendragon to have the Round Table constructed, on the model of the Grail table and with a place for the future winner of the Grail. That the magic circle of the Round Table should be designed by a master magician is suitable enough, but the effect is to give Merlin a substantially more dominating position than before. He was already the genius who brought Arthur on to the stage of history. Now he inspires and prompts the whole of Arthurian knighthood, its highest ideals and the quest of the Grail. It is made clear that Merlin does not cause the events which are to occur, but it is through him that the secret forces of destiny penetrate the human world. It is not the Christian Church, as one might expect, which carries out this task but the great enchanter, who is half a demon and has his roots in the pagan past. The implication is that the quest of the Grail is something older, stranger and more profound than the Church can contain.

Robert de Boron's story was continued by other authors (principally in the Vulgate Cycle and the *Suite du Merlin*). Merlin was a useful character from a story-teller's point of view, because his knowledge of the future enabled him to make telling remarks, which registered with the reader if not with the other characters. In the sequels he continues to act as Arthur's mentor and almost as a magical substitute for a father confessor, guiding the young king through the early hazards of his reign. He uses his magic to protect Arthur from danger; he takes him to the lake to obtain Excalibur; he goes to ask for Guinevere's hand for Arthur; he decides which knights are to be appointed to the Round Table.

It is not uncommon in hero legends for the central character to have a supernatural helper, who may be male or female. If male, he is frequently a magician, a hermit or a craftsman. Merlin combines all three and the romances seem to have been responding to a general human instinct in making him the guide, philosopher and friend of Arthur and the Round Table. The helper is often an inscrutable and sinister being as well as a benevolent one, perhaps because he or she leads the hero into the mysterious realm, the otherworld or the depths of the mind, where he encounters forces which are both fruitful and dangerous. Merlin's role in paving the way for the quest of the Grail seems to fit in with this.

Merlin was a craftsman as well as a magician, or rather the two roles were not clearly distinguished. An example is the engineering skill which he brought to bear on Stonehenge. Another is the tale of him constructing, 'by his subtle craft', a magnificent tomb for King Loth, Gawain's father. It had figures of the kings slain with Loth in battle, each holding a lighted wax taper, which never burned out, and an image of Arthur with drawn sword standing above them.

Shape-shifting was another of Merlin's magical abilities. He appeared in many different forms, among them an old man with a sickle, a boy of fourteen, a cripple, a beggar, a hermit, a woodcutter, and a shadow. He could also transform other people's appearance, as in Uther's case. The belief that a master magician can change shape goes far back beyond the Arthurian legends to the shamans of prehistoric tribes in Europe. Because the magician can take any form he chooses, he can adopt the identity, share the experience and thereby understand the nature of any creature. In this way he comes to comprehend the realities behind appearances.

In Celtic tradition the Druids were believed to have this ability, and so were great poets, it being the business of a poet to penetrate the nature of all phenomena, human and other. The Myrddin of the original Welsh tradition was a bard.

If a magician can change bodies, it becomes doubtful whether he ever really dies, and there are many different accounts of what became of Merlin. Some of them have him carried off to hell by the demons who sent him into the world in the first place, but in most of them he does not exactly die but disappears into a state of suspended animation: like Arthur resting in Avalon or sleeping in a cave, but without the promise that he will return. The line in the 'Afallenau', 'Death has taken everyone, why does it not call me?', may contain the germ of the idea that Merlin was unable to die.

A late story says that, taking the Thirteen Treasures of Britain with him, Merlin vanished into the House of Glass on Bardsey Island, off the Caernarvonshire coast of Wales, and there he remains to this day, concealed from mortal eyes. According to a thirteenth-century story (in the Didot *Perceval*), he survived Arthur and all the knights of the Round Table. When Arthur was felled in the final battle against Mordred and departed to Avalon, Merlin retreated into what he called his *esplumeor* and was never seen again. An *esplumeor* was a cage in which birds were kept during the moulting season so that their feathers could be collected. This story may imply that Merlin ended his days caged like a bird, or that he experienced a transformation and renewal at the beginning of another phase of his existence, like a bird shedding its worn-out feathers and growing new ones.

More often, Merlin vanishes from the scene much earlier. The most popular story, of which there are several variants, was that he fell in thrall to a beautiful maiden or fay named Vivien (or Niviene or Niniane), who was sometimes identified as the Lady of the Lake, later the foster-mother of Lancelot. Playing on his desire for her, she turned his own magic against him and shut him up in a cave or a tomb or a castle of air, where he remained a helpless prisoner.

Earlier episodes imply that for all his craft and sagacity Merlin was helpless against passionate love. He could not prevent Arthur's incest with his half-sister. When Arthur fell in love with Guinevere, Merlin foresaw the tragic consequences but all he could say to Arthur, in effect, was: 'If you have set your heart on her, I cannot dissuade you.' Merlin himself hankered vainly for Morgan le Fay, who learned some of her skill in magic from him. Finally his doom overtook him because he could not resist the charms of Vivien. Unfortunately for Merlin, Vivien was not interested in him for his own sake but only for the sake of his magic, and indeed there are few records of his being loved by a woman. Perhaps this is because he was only half human and his human half was female; perhaps it is because his male nature was demonic, which made him capable of lust but not of love.

According to the Vulgate *Merlin*, the enchanter first saw Vivien by chance at a spring in Brittany. Smitten with her at once, he took the form of a handsome young man, told her he was a magician and created enticing illusions out of thin air to impress her: a cavalcade of knights and ladies, a lovely garden, young knights jousting, minstrels singing. Vivien promised to become his mistress if he would teach her the secrets of his magic and he readily agreed to the bargain. Coaxing him and leading him on without ever satisfying his desire, she learned more and more from the besotted enchanter until finally she discovered the last secret she needed. One day when they were in the enchanted forest of Broceliande, sitting beneath a whitethorn bush in full blossom, she lulled him to sleep with his head in her lap. Then she drew a magic circle nine times round him with her girdle and when he woke up he found himself in a tower built of mist. She kept him there as her prisoner and though she went in and out as she liked, Merlin could not escape.

Because Merlin could foresee the future, he knew perfectly well what was going to happen, but he could not help himself. Caught in the net of an irresistible passion, his struggles grew gradually weaker. The detail of Vivien lulling him to sleep with his head in her lap recalls the legend of the unicorn, which is tamed when it bows its head with the single phallic horn and rests it in the lap of a maiden. It is the sign of Merlin's total sexual subjugation to Vivien and the final loss of any will to resist her. The girdle with which she draws the magic circle round him is the symbol of her sex. The tower of mist is an otherworld castle and the whitehorn bush may be descended from the white otherworld boughs of Celtic tradition which charmed mortals to sleep. The spectacle of the great magician hoist with his own petard, the builder of illusions finally trapped in an illusion, makes an ironic conclusion to his career.

The *Suite du Merlin* has a more elaborate version of the story. At the wedding feast of Arthur and Guinevere, a white stag burst in and careered through the hall, pursued by a pack of hounds in full cry and a beautiful, boisterous huntress in a green robe, brandishing a bow. She was the Lady of the Lake (and for reasons which will become apparent is here given the attributes of the Roman goddess Diana). She stayed at Camelot as a guest and Merlin desired her. She valued her virginity and detested Merlin, but by pretending to enjoy his company she wormed the techniques of his magic out of him.

When Vivien left Camelot to return home, Merlin insisted on accompanying her. In the land of Benoic they saw the baby Lancelot, and Merlin predicted that within twenty years the child would be the most honoured man in the world. They went on, until they came to a lake in a wood. This was the Lake of Diana and Merlin explained how it had come by the name. In the time of Virgil a beautiful huntress named Diana lived by the lake with her lover, Prince Faunus. Diana grew tired of Faunus and fell in love with another man. She decided to rid herself of Faunus and, persuading him to lie down in a marble tomb, she murdered him by pouring boiling lead over him. When the crime was discovered, she was beheaded and her body was thrown into the lake, which was named after her.

Vivien coaxed Merlin into building an enchanted, invisible palace by the lake and they stayed there for a time, but as his craving for her grew, so did her secret loathing of him. Merlin told her that Arthur was in mortal danger from Morgan le Fay but he hesitated to go to help Arthur because he knew that his own death awaited him if he did. Merlin could not see clearly what his fate would be, because his gift of foresight had been weakened by the spells which Vivien had learned from him and was using against him. She told him that it was his duty to return to Britain and she promised to protect him from all harm.

They set out together and presently came to the Perilous Forest, where they found among the rocks a tomb in which two unhappy lovers had been buried. Taking a leaf out of Diana's book, Vivien wheedled Merlin into removing the heavy marble lid of the tomb.

Then she cast him into an enchanted sleep. Summoning her servants, she told them that the magician was now himself overcome by magic because of his lustful intentions towards her. They put Merlin into the tomb with the bodies of the lovers and covered it again, and Vivien sealed it with magical conjurations so that it could never be opened. She left him there to endure a living death.

According to a Spanish version of the story (*Baladro del Sabio Merlin*), three days after Vivien had shut Merlin in the tomb, one of Arthur's knights came riding by and heard the voice of Merlin, bewailing his fate. The knight said he would set the magician free at once, but Merlin told him that it was hopeless, for the enchantment could not be broken. Not long before noon the knight heard Merlin calling out in a demonic voice for his father to come for him. A heavy darkness gathered, with a rising storm and a gabble of eerie voices, and Merlin uttered one last terrible cry. The cry was heard three leagues away and at that moment in Camelot the wax tapers burning in the hands of the figures on Loth's tomb suddenly went out. This description makes Merlin's final moments an evil mimicry and counterpart of the death of Christ.

Two well-known themes are combined in the story of Merlin and Vivien. One is the tale of an enchantress or fay who lures a man to her otherworld home and holds him prisoner there. The other is the sardonic theme of the wise man whose sagacity is not proof against sex and whose itch for a woman much his inferior in attainments turns him into her humiliated slave. There are probably echoes from the Welsh legend of Myrddin the wild man in the story. Wild men of the woods in medieval folklore were notorious for their amorousness. Apple trees, which are mentioned prominently in accounts of Myrddin's life in the woods, were otherworld trees linked with female sexuality. The lines in the 'Afallenau' about Myrddin's dalliance with a maiden at the foot of an apple tree may have influenced the tale of Merlin held in thrall by Vivien. His forest house with seventy doors and seventy windows summons up a picture of something not too far removed from a castle of air or the House of Glass on Bardsey Island. In the *Suite* Merlin plays the role of Faunus, and Faunus was originally a Roman spirit of the uncanny woods.

The episode of the visit to Lancelot's cradle supplies an explanation for the Lady of the Lake's subsequent interest in the hero, and one of Diana's functions in Roman belief was the care of new-born children. Assimilating Vivien to Diana was a shrewd stroke. In the early middle ages the Lady Diana was associated with witchcraft. It was widely believed that witches rode out with her at dead of night, and so she was a suitable figure to connect with an enchantress. Diana, in classical mythology, was a virgin goddess. Her principal pleasure was hunting, and she was so prudish that in a well-known story she punished Actaeon, for seeing her naked, with death at the jaws of her hounds. The hearty, tomboyish character assigned to Vivien through this association seems well calculated to appeal to Merlin, whose human side was feminine, and also explains her horror of his advances.

In a final brilliant rounding-off of the story in the *Suite du Merlin*, Vivien takes Merlin's place as Arthur's supernatural helper. The Lady of the Lake has robbed Arthur of his wisest counsellor but it is she, using the magic she has learned from Merlin, who saves him from the stratagems of Morgan le Fay.

Morgan le Fay is the feminine counterpart and opposite of Merlin. Where Merlin brings Arthur into the world at the beginning of the story, Morgan takes him away from the world at the end. Merlin loves and helps Arthur, Morgan hates and harms him. In the earliest stories about her which have survived, however, Morgan has a benevolent role. She is the ruler of the Isle of Avalon and it is to her that Arthur is taken to be healed after his last battle.

In this capacity, Morgan comes from the seam of Celtic myths about otherworld islands of women. An example of the genre is the Irish story of Bran son of Febal, who sailed to the west with his men to find the island of Emain Ablach, or Emain of the Apple Trees. When they reached it they were welcomed by the queen and royally entertained, with luscious women to share their beds, and food which did not diminish when they ate it. The island stood on four columns of white bronze which rose out of the sea. There was no death or decay there, no sickness or sorrow, nothing to mar perpetual joy in feasting and lovemaking, music and wine, jewels, birds, flowers and trees, horses which were golden and blue and purple, and chariots of silver and gold. Bran and his men stayed there for a year and then one of them grew homesick and they decided to take him back to Ireland. The queen warned them not to go and, when they insisted, told them on no account to touch the land. When they came close to the Irish coast they hailed the people there and discovered that they had been away for not one year but many hundreds of years. They were already a legend. The homesick man jumped out of the boat and the moment he touched the ground he turned into a pile of ashes, like one long dead. Bran and the rest sailed away, and no one knows what became of them.

There is another story about Conle the Redhaired, who was a son of Conn of the Hundred Battles. A beautiful fay from a fairy hill fell in love with him and gave him an apple. For a month he lived on nothing except the apple, which stayed the same size however much he ate of it, and a powerful longing for the fay possessed him. She came in a crystal boat and called to him. He struggled against his need for her, because he loved his own people, but he could not resist. He sprang into the boat, sailed away with her, and was never seen again.

Both these stories are tragic. The hero goes to a paradise of erotic and sensuous pleasure which will last for ever, but he cannot return. Though he seems to have gained everything desirable, he has lost the capacity for action which makes life worth living. Helpless in a beautiful dream, he is no longer truly a man. There is a parallel in the Greek story of Circe, who lured men to her lovely island home and turned them into animals. In other words, she robbed them of their human nature. This aspect of the otherworld helps to explain the alluring but sinister nature of Morgan and other Arthurian fays and enchantresses.

The Celtic tradition of uncanny islands was not only a matter of myth. In the first century BC the philosopher Posidonius reported that an island off the mouth of the Loire was inhabited by a female religious community. Once a year they put a new roof on the temple of their god. The work had to be done in a single day, and if one of the women accidentally dropped any of the roofing material the others tore her to pieces and carried her remains round and round the temple. No males were allowed on the island, but the women occasionally left it and returned to the mainland to sleep with men. Customs of this sort were probably bound up with the tales of islands of women and the theme of the amorous fay who comes from her otherworld home to seek the love of a mortal.

According to Geoffrey of Monmouth, Morgan had healing powers and the ability to fly, and lived with her eight sisters on the Isle of

Avalon. Nine, which is three times three, was a sacred number to the Celts. The story may go back to a real community of women, of the type recorded by a Roman writer of the first century AD, Pomponius Mela. He said that nine virgin priestesses lived on the Ile de Sein, off the Breton coast, who were believed to have magical powers. They could cure the sick, control the weather and the sea, foretell the future and turn themselves into animals.

The fays of legend seem to have mingled traditions of powerful priestesses of this kind with memories of the great goddesses of Celtic paganism. Several medieval authors referred to Morgan as a goddess, and it is likely that she was descended from a trio of Irish goddesses of love and war, known as the Morrigan (great queen), which was apparently the name of one of these goddesses as well as of the three together as a group. Celtic mother goddesses had equal powers of creation and destruction, of giving life and taking it. The Morrigan croaked and strutted about battlefields in the form of ravens and crows to feast on the dead. She had an insatiable sexual appetite and could appear as a seductively beautiful woman, but she also manifested herself as a hideous hag, inspiring terror and revulsion.

The Morrigan desired the hero Cuchulain and came to him as a lovely girl in a dress of many colours. He put her off with the excuse that he was too busy with fighting and too exhausted for lovemaking, and then she offered him her help. When he scorned the help of a woman, she was angry and told him she would attack him when he was in serious danger. This she did, choosing a time when he was fighting a formidable adversary at a ford. She appeared first as a herd of white cows which charged him, next as a huge black eel which coiled round him, and then as a she-wolf. Cuchulain drove her off each time. She retained a strong affection for him, however, and when he fought his last battle she fluttered about him in the form of a crow, in anxious distress, until his enemies overpowered him and killed him. Then she flew away.

This story about Cuchulain and the Morrigan turns on the same tension between action and love, the same temptation to forsake the field of glory and find oblivion in a woman's arms, which is the theme of many Arthurian tales. The Morrigan's relationship to Cuchulain resembles the love–hate relationship between Arthur and Morgan, who hates Arthur and tries to kill him, but who later saves him from death and takes him to live immortally with her in Avalon. No satisfactory explanation for Morgan's hatred of Arthur is given in any of the surviving stories. Partly, at least, Morgan stands for feminine resentment of men, who lead careers of action in which women are not allowed to share (a resentment which lies behind the enthusiasm with which the troubadour ideal of romantic love was taken up by gifted, highly educated women like the Countess Marie of Champagne, Chrétien de Troyes' patroness). There may once have been a story that Morgan desired Arthur and he rejected her. If so, this would help to explain the relentless vendetta which the enchantress waged not only against Arthur but against Guinevere, his wife. Not until the end of Arthur's career on earth was Morgan able to claim him for her own.

Morgan's connection with Avalon probably comes from another strand in her ancestry, her descent from the Welsh goddess Modron, 'the Mother'. One of the triads says that Modron was the daughter of Avallach, the wife of Urien and the mother of Owain. In the Arthurian stories Morgan is the wife of Urien and the mother of Yvain, Owain's French equivalent. Morgan has taken Modron's place, apparently because the Irish Morrigan and the Welsh Modron had coalesced. It seems that Avalon was originally the island of the Welsh god Avallach, which Modron or Morgan inherited as his daughter. There was also a tradition of another male ruler of Avalon. In Chrétien de Troyes his name is Guingamars, and Morgan is his mistress. He turns up again in Malory as the lord of Avalon and the brother of Lady Lyonesse. Usually, however, Morgan is the lady of Avalon in her own right.

One mark of Morgan's descent from Irish and Welsh triple goddesses may be her frequent appearances as the leader of a group of three fays. Another is her ability to fly, for the Morrigan, and probably Modron too, could take the form of a bird. Morgan also inherited the goddess's dual nature as a beauty and a hag. In most stories Morgan is beautiful and desirable. In some she is an ugly old crone. Christian influence produced the explanation that the fay became prematurely old and hideous because of her addiction to evil magic, but there are hints here and there of the earlier pagan conception of the goddess who is sometimes young and lovely, sometimes withered and ugly. In *Sir Gawain and the Green Knight* Morgan is the old woman in the green knight's castle who has planned the whole enchantment as a plot against the Round Table. She is a squat and wrinkled hag, fat-bottomed, black-browed and bleary-eyed. The green knight's wife, who is Morgan's agent and tempts Gawain, may be Morgan's other and younger self as a bewitching seductress.

The evil side of Morgan's character is generally uppermost. She is Arthur's sister, or more often his half-sister, the youngest of Ygraine's daughters by her first husband. She was sent to school in a nunnery, and Malory liked to think that she had learned her mastery of black magic there, but the usual story was that Merlin taught her the arts of enchantment. Morgan was married to King Urien of Gorre. The name of his country, as in Chrétien's *Lancelot*, suggests the otherworld, the Ile de Voirre or Isle of Glass. She took numerous lovers and eventually she tried to kill Urien, prevented only at the last moment by their son, Yvain. She also attempted to murder Arthur, in the affair of the false Excalibur.

The tradition that Morgan was the mistress of Guingamars, lord of Avalon, was used as the basis for a story intended to account for the fay's savage hostility to Guinevere. Morgan's first love was a young knight of Arthur's court named Guiomar (derived from Guingamars). At this time she was one of Guinevere's ladies-in-waiting. Guinevere caught the lovers together and, anxious to avoid a scandal in the royal household, persuaded Guiomar to give Morgan up. Morgan never forgave Guinevere and longed for revenge on her ever afterwards. She left court and went to find Merlin, offering herself to him if he would teach her magic. Merlin agreed, but Morgan did not keep her promise.

Her experience had made Morgan cynical about love and lovers. Using the arts she had learned from Merlin, she created a green and delightful valley, watered by a sparkling spring and enclosed by a wall of air. It was called the Valley of No Return, because no knight who had been untrue in love could escape from it. Morgan soon trapped the faithless Guiomar there, and many other knights. Her prisoners passed the time happily in feasting and dancing and, if they had their ladies with them, in enjoying their caresses. The valley's spell could be broken only by a knight who had never been false to his love, even in his thoughts.

It was Lancelot who broke the enchantment. He rode into the valley and came to a narrow place on the road, guarded by two dragons. He killed the dragons, crossed a turbulent river bridged by a narrow plank, defeated five of Morgan's knights who defended the crossing, and pressed on through a wall of flame to Morgan's palace. He routed two more knights and chased a third through a succession of rooms and across a garden into a luxurious tent, where the fay was lying asleep on a couch. The fleeing knight tried to hide under the couch, but Lancelot routed him out, tumbling Morgan out of bed in the process, and cut his head off. He apologized to Morgan for the disturbance and politely offered her the severed head. She hid her fury as best she could and the knights she had held prisoner came and hailed Lancelot as their rescuer.

The Valley of No Return is another of the otherworld realms in which men of action are trapped in a helpless daze of sensuality. In the background is the idea of the land of death, a country from which no traveller returns, and the hero who invades the otherworld and sets the prisoners free achieves, in essence, a victory over death. Morgan does not inflict physical hardship or torture on her victims in the valley. Their punishment is a deprivation of the spirit, exile from the life of action. For them it is a living death. From the feminine point of view, or so the story implies, the situation is quite different. At the end, when the enchantment is broken, the captive knights are delighted to be released, but their mistresses are not. The ladies have been thoroughly content in the spellbound valley. They have enjoyed the undivided attention of their lovers, for once undistracted by the demands of adventure and fame. They resent the ending of the idyll and the return to real life.

Morgan frequently wove her webs for Lancelot himself, as we have seen, and there are several other stories of her luring knights to her enchanted realm. She had designs on a promising young knight named Alexander the Orphan and after he had been badly wounded in a joust she carried him off to her castle of Belle Garde (Beautiful Keep). She told him she would heal his wounds if he promised to stay there as her lover for a year and a day (in Celtic tradition, the usual interval between a betrothal and a wedding). Alexander was in searing pain and reluctantly gave his promise, but as his wounds mended he resented his servitude. He was rescued by another woman, who came to visit the fay. She took pity on Alexander and contrived to burn the castle down. Morgan fled, but Alexander stayed among the ruins until the year and day were up, in order not to break his word.

If Merlin is Morgan's male counterpart, her female one is the Lady of the Lake. Both are beautiful fays, enticing and dangerous. Both are taught magic by Merlin, as part of a bargain which neither of them keeps, but the Lady of the Lake uses her magic arts to protect Arthur and Lancelot, while Morgan uses hers against them. And yet, in the end, it is Morgan who comes to Arthur when he is mortally wounded and heals his wounds in Avalon. Morgan's powers of healing, which are frequently mentioned, even in stories in which she is an evil sorceress, show that there was originally a benevolent and protective side to her character. Christian writers may have found it hard to cope with such an ambivalent figure (as some of them found the ambivalent Merlin difficult to handle). It may be that the two sides of Morgan's nature separated into two different characters and that the Lady of the Lake is an aspect of Morgan herself. If so, the two fays represent the two aspects of the great goddess, who is fertile and destructive, motherly and murderous, loving and cruel. It was perhaps with a sure instinct that Malory made both Morgan and the Lady of the Lake come to the dying Arthur after his last battle and take him away to Avalon.

However this may be, in most of the stories about her Morgan personifies the old and deep-rooted male fear of the evilness of woman, which is not confined to the Celtic background but exists in the Judaeo-Christian and classical traditions, to which Arthurian writers were also heir. Woman's evilness is linked with voracious female sexuality, felt to rob man of his dominance and reduce him to abject subjection. It is connected with an old awareness of the irrational and overwhelming nature of passionate desire, regarded as a supernatural force. Many adjectives commonly applied to desirable women – enchanting, entrancing, fascinating, bewitching, glamorous, charming – have lost their original meaning now, but in the past they implied the exercise of an alluring but sinister magic, a hypnotic influence projected by a beautiful woman which put her victims at her mercy. There is an obvious symbolic link between the female genitals and the persistent notion of the enchantress whose lair is in an enclosed place – a castle, a garden, a secluded valley – or by a spring in the recesses of a forest: though there is nothing to show that medieval writers were conscious of it.

Beyond this, Morgan and the other Arthurian fays and enchantresses stand not only for woman as a honeyed but deadly trap, but more generally for the powerful attraction of the desire to escape from real life. Their enchanted castles and palaces, their valleys and gardens, are retreats from reality. They are places where no changes occur, no decisions are made, no battles are fought, no challenges are faced. They hold out an illusory promise of ease and peace and pleasure. Ultimately the beautiful fay stands for something in the hero's own nature, the longing which now and again besets him to abandon his quest, to give up the struggle and pursue a false dream of happiness gained through luxury, sensuality and ease.

3
The Quest of the Grail

> But always in the quiet house I heard,
> Clear as a lark, high o'er me as a lark,
> A sweet voice singing in the topmost tower
> To the eastward: up I climbed a thousand steps
> With pain: as in a dream I seemed to climb
> For ever: at the last I reached a door,
> A light was in the crannies, and I heard,
> 'Glory and joy and honour to our Lord
> And to the Holy Vessel of the Grail.'
> Then in my madness I essayed the door;
> It gave; and thro' a stormy glare, a heat
> As from a seventimes-heated furnace, I,
> Blasted and burnt, and blinded as I was,
> With such a fierceness that I swooned away –
> O, yet methought I saw the Holy Grail,
> All palled in crimson samite, and around
> Great angels, awful shapes, and wings and eyes.
>
> Tennyson, *The Holy Grail*

The legends of the Grail have an enthralling atmosphere of mystery, of some tremendous secret which stays tantalizingly just outside the mind's grasp, in the shadows beyond the edge of conscious awareness. The outlines of the secret become clearer as writer after writer takes up the theme and makes his own sense of it, but we are never told in plain language exactly what the Grail means. It is 'the perfection of Paradise' and its presence brings an 'abundance of the sweetness of the world'. It is 'the source of valour undismayed' and 'the spring-head of endeavour'. The inner mystery of the Grail cannot be explained, because it is 'that which the heart of man cannot conceive nor tongue relate'.

When the Grail appears, in the stories, some people see it but others may not. Those who see it, see it in different forms. It is kept hidden in a castle, but it can show itself elsewhere. It may be carried by a beautiful woman or it may move about by itself. Because it is a holy thing, it is uncanny and dangerous. The story-tellers speak of it in hushed voices. Nobody can talk of the Grail, they say, without trembling and turning pale. Anyone who tells its story falsely will suffer harm and pain. Its secret should never be revealed because, before the tale is fully told, something may be stirred up that is better left unaroused.

On one level, obviously, this is an artistic device to whet the audience's appetite, but these comments by medieval writers suggest that they were dealing with a mystery which they recognized as both profound and unorthodox. The Church, far from taking the Grail under its wing, treated the legends with cold reserve.

In the earliest account of it that has survived, the Grail seems to be a Christian talisman, or power-containing object, a mysterious manifestation of God. In the later stories its Christian origins and significance are elucidated. Usually it is the chalice which Jesus used at the Last Supper, when he took a cup of wine and gave it to the disciples, saying: 'This is my blood of the covenant, which is poured out for many.'[1] It was believed that some of the blood which flowed from the Saviour's wounds on the cross was collected in the same cup. In an age which eagerly believed in the miraculous power of sacred relics, the idea that the cup of the Last Supper and the divine blood had survived, guarded somewhere in secret, caused profound awe and excitement. Malory describes the coming of the Grail to Camelot, when Arthur and his knights sat down to supper at the Round Table on Whit Sunday:

> Then anon they heard cracking and crying of thunder, that the palace should all to-drive [burst to pieces]. So in the midst of the blast entered a sunbeam, more clearer by seven times than ever they saw day, and all they were alighted of [illumined by] the grace of the Holy Ghost. Then began every knight to behold other, and every saw other, by their seeming, fairer than ever they were before. Not for then there was no knight that might speak one word a great while, and so they looked every man on other as they had been dumb.

Then entered into the hall the Holy Grail covered with white samite, but there was none that might see it nor whom that bare it. And there was all the hall fulfilled with good odours and every knight had such meats and drinks as he best loved in this world.

And when the Holy Grail had been borne through the hall, then the holy vessel departed suddenly, that they wist not where it became. Then had they all breath to speak, and then the king yielded thankings to God of His good grace that He had sent them.[2]

The sacred relic was called the *Saint Graal* (Holy Grail), which was later elided to *Sangraal*, which through an ingenious pun turned into *Sang Réal*, 'Royal Blood', the blood of Christ the King. But the Grail was not always the cup of the Last Supper and the blood of the Redeemer. It might be the dish from which Jesus and his disciples ate the Passover lamb at the Last Supper. It might be a reliquary or a ciborium, which was a cup with a lid topped by a cross, in which the consecrated hosts or Mass-wafers were kept. The Grail might even be a mysterious stone which had been brought down to earth from heaven by angels, and in one eccentric story it is a platter containing a severed human head swimming in blood.

It seems clear that although the Grail is a Christian object, there is something much older behind it. The origins of the Grail legends have been searched for, and duly found, in Christianity, in Celtic paganism, in the classical mystery religions and among Byzantine, Persian, Jewish and Islamic traditions. Apparent traces of gnostic ideas have been discovered and supposed signs of heretical influences. There have been enticing but probably ill-founded suggestions of secret cults, whose ceremonies and symbols can be discerned in the tales. So many different theories are possible because the stories that have come down to us vary very considerably. What the Grail is, what it looks like, what it does, who guards it, what the purpose of the quest is, who achieves the quest, how he achieves it and what happens then: on these and innumerable other points there is remarkably little agreement between the different tales. On the contrary, the great American authority R. S. Loomis, who believed that the sources of the Grail legends were primarily Celtic, complained that 'the authors of the Grail texts seem to delight in contradicting each other on the most important points'.[3]

The probability is that the Grail legends were drawn initially from the same reservoir of Celtic tradition as the rest of the Matter of

Britain. There were old Celtic stories of magic vessels of life and regeneration in the otherworld, seen and sometimes carried off by heroes. Medieval writers took up this theme and put it into their own Christian framework because they sensed in it an obscure but profound significance, closely related to the theme of the hero's quest for integrity and his victory over death. The magic vessels were the ancestors of the Grail. In one line of pagan tradition, the otherworld vessel was connected with the succession to a kingdom, and so other themes were added to the central one. One of them was a tradition, or a group of traditions, about a king who was crippled or feeble with age and as a result his land had fallen desolate or was under a spell. The hero, who was the king's heir, had to break the spell. Another was a story of revenge, in which the hero proved his right to the throne by avenging a murdered kinsman. Mingled with these was the idea of an immensely destructive otherworld weapon, a spear or a sword.

Different authors put these building-blocks together in varying ways in the light of their own skill and insight, and this is why they contradict each other so frequently. They took various disconnected or loosely connected pagan themes and fused them into more or less coherent Christian stories. On the whole, the stories grew more coherent as time went by, inconsistencies were ironed out and the building-blocks were more efficiently assembled. The point is important because a good deal of writing about the Grail has been based on the opposite assumption, that what we have are the fragments of a coherent pagan myth, muddled up by medieval story-tellers.

What we really have are stages in the making of a Christian myth. The heart of the myth is the idea of an object of awesome sanctity and power, which holds the secret of life. The stories, for all their diversity, are about a hero winning his way to this object, making himself master of it and penetrating its mystery.

The Fisher King's Castle

The earliest surviving story about the Grail is the *Conte du Graal* (Story of the Grail), written by Chrétien de Troyes in about 1180. Unfortunately, he died before he could finish it. The hero is Perceval, a young man who lived with his mother in the depths of a wild forest in Wales, near Snowdon. His father had retreated to a quiet manor-house in the forest years before, after being crippled in battle by a wound in the thighs, and had died when Perceval was a small boy. Perceval was brought up by his mother, who deliberately kept him in ignorance of the great world beyond the forest's edge. He had never heard of knights and chivalry, or of Christ and the Church, though he had been told a little about God and the angels and demons. He had no notion of civilized manners and sex was a closed book to him. He did not know his own name and his mother called him Fair Son. Handsome, sturdy and engagingly cheerful, he was impetuous, brash and comically naïve.

One day in the forest Perceval saw five knights in all their finery of glittering armour and glowing heraldry. He thought they must be angels and that their leader was God. They disillusioned him, not too unkindly, and explained that they were knights of King Arthur's court. Perceval at once decided to be a knight himself. His mother did her tearful best to dissuade him, but he brushed her protests aside and in the end she sent him out into the world with a clumsy, rustic suit of clothes and some hasty instructions about women, churches and the fundamentals of Christianity. When he left home, his mother fainted with grief. He looked back and saw her lying on the ground as if dead, whipped up his horse and rode away.

Perceval found Arthur at Carlisle, rode into the hall and demanded to be made a knight. The king did not think him ready for knighthood yet, though a girl at the court predicted that if he lived long enough he would be the best knight in the world. Perceval hurried away to prove himself. He was a redoubtable fighter, but so simple-minded and uncouth that people who encountered him were not sure if he was quite right in the head. After a time, however, he came to the castle of a lord named Gornemant, who knighted him after teaching him the code of chivalry and the correct management of lance, shield and warhorse. Gornemant also told Perceval to curb his tendency to chatter and to say whatever came into his head. Talking too much, Gornemant said, was sinful and rude.

Now that he was a knight, Perceval meant to return to his mother, for he was worried about her, but he was delayed by a series of adventures and by falling in love with a delightful girl named Blancheflor. Still meaning to go home, he came one day to a river, where he saw a small boat with two men in it, one of them fishing. The fisherman told him there was no way of crossing the river for miles and offered him hospitality for the night at his own castle nearby, in a valley beside the river.

Perceval went to the castle, where he was graciously received by its lord, the fisherman, who was crippled and reclined on a couch in front of a blazing fire. He presented Perceval with a superb sword, telling him that it was destined for him. A young man came into the hall carrying a white lance by its shaft. From its iron head a drop of red blood ran down on to his hand. Perceval was naturally curious about it, but he remembered Gornemant's advice to be reticent and he kept quiet. Two more young men came in, carrying golden candelabra with many burning candles, and with them a lovely girl, richly dressed, who held in her hands a golden grail, studded with precious stones finer than any that could be found in the earth or in the sea. When she came in with the Grail, there was such a radiance of light that the candles lost their brightness, like stars dimmed by the sun or the moon. After her came another damsel, carrying a silver carving-dish. The procession crossed the hall and disappeared into a room at one side.

For the same reason as before, Perceval did not ask about the Grail or enquire who was served from it. A sumptuous dinner was brought to Perceval and his host, and as each course was served the Grail passed before them. Still Perceval politely said nothing, though he was consumed with curiosity and decided to ask one of the young men about it the next day. After dinner the lord and Perceval passed the rest of the evening talking. The lord was carried off to bed in a litter and Perceval bedded down in the hall. When he woke up the next morning, there was no sign of anyone in the castle and the doors to the rooms off the hall were locked. Taking his new sword, Perceval mounted his horse and rode out over the drawbridge, which began to rise under him so that his horse had to jump to the far side. He called back to ask who had raised it, but there was no answer.

Perceval rode away into the forest. There he met a girl who was sitting under a tree, lamenting and cradling in her arms the headless corpse of a knight, her lover. The girl proved to be Perceval's cousin. She told him that he had spent the night at the castle of the rich Fisher King, who had been wounded in battle by a javelin thrust through both his thighs. In constant pain, he could not walk or mount a horse and so he amused himself with fishing. Had Perceval asked why the lance bled? No? Then he had done badly. Had he asked where the Grail procession was going? No? Then he had done worse.

At this point the girl asked Perceval who he was, and suddenly he knew his name. He told her he was Perceval of Wales and she said he ought to be called Perceval the Unfortunate. If he had asked the right questions about the lance and the Grail, he would have healed the Fisher King and great good would have come of it. Now there was misery in store for Perceval himself and for others. He had failed because of his sin against his mother, who had died of sorrow when he left her. The sword which the Fisher King had given him would betray him at a critical moment and break to pieces. It could be repaired only by a smith named Trebuchet at the lake near the Firth of Forth.

After this, Perceval made his way to Arthur's court, which he found at Caerleon. He was welcomed joyfully as a hero because of his exploits while away, but the rejoicing was soon stilled. A hideously ugly maiden rode into the king's hall on a tawny mule. She was crooked and hunchbacked, with eyes like a rat, a nose like a cat, lips like an ass, yellow teeth and a beard like a goat. Her neck and hands were black. Her black hair was tied in two braids and in her right hand she carried a scourge. Sitting on the mule, she greeted Arthur and his lords, and then berated Perceval for failing to seize his opportunities at the Fisher King's castle. Because he had not asked about the lance and the Grail, the Fisher King would never hold his estates in peace. Lands would be laid waste, knights killed, women widowed and children orphaned because of Perceval's silence. The ugly maiden rode away. Perceval vowed never to spend two nights in the same place or shirk any danger or suffering until he discovered why the lance bled and whom the Grail served.

The story now turns away from Perceval to follow the adventures of Gawain. When we return to Perceval, five years have gone by. In that time he has survived many perils and defeated many valiant knights, but he has never entered a church or worshipped God or even thought of God. He has also come no nearer to the object of his quest. On Good Friday he meets knights and ladies who reproach him for riding in arms on the day Christ died. Puzzled and moved, Perceval consults a hermit, who turns out to be his uncle, his mother's brother, and to whom he tells his whole story. The hermit explains that all Perceval's troubles stem from the grief he inflicted on his mother, which caused her death. This sin tied his tongue in the Fisher King's hall. There are two kings in the Rich Fisher's castle, the Fisher himself and his father, who is another of Perceval's maternal uncles. It is the old king who is served by the Grail. What it serves him, the hermit says, is not a pike or a salmon but a single Mass-wafer. The Grail is so holy and the old man so spiritual that this is all he needs to sustain him, and he has lived on nothing else for fifteen years. During that time he has never left his room in the castle.

The hermit instructs Perceval to love and worship God, revere the priesthood, honour good men and women, and help ladies in distress. He teaches Perceval a prayer, full of names of Christ which are of such power that the prayer should never be uttered except in peril of death. The wording of this prayer is not revealed to us. The story now returns to Gawain and deals with his exploits at the Castle of Wonders and his love for the beautiful Orgueilleuse. Chrétien never finished it, and we hear no more of Perceval.

Chrétien described the story of the Grail as the best tale that could be told in a royal court. He wrote it at the request of Count Philip of Flanders and based it on a book which Count Philip lent him, so whatever Chrétien may have made of it, he did not invent the Grail. There is a suspicion that the episode about the hermit was added by someone else later on. The story is consequently the work of two or perhaps three different authors: the unknown author of Count Philip's book, Chrétien himself, and possibly another unknown writer who supplied the hermit episode in an attempt to explain the Grail.

Taking the story of Perceval as we have it, however, its central theme is the hero's progress towards an ideal integrity as the best knight in the world, given spice by the fact that he is such a singularly unpromising candidate to begin with. The tale opens in spring, the right season for the start of a new life. Perceval is an overgrown child, winning but naïve and boorish, kept safe by his mother in the nursery of the sheltering forest, which cuts him off from the real world. When he sees the five knights, he senses the way to manhood. Determined to escape from the nursery, he is too self-centred to care about hurting his mother. He breaks into the real world and the next important stage in his development comes when he learns the code of chivalry with its emphasis on duty to others. He is gradually turning into a civilized human being instead of a bumpkin, and he now becomes worried about his mother. Once he is knighted, all his exploits are undertaken to help women in distress and he is exposed to the refining influence of love by the beautiful Blancheflor.

Perceval still has a long way to go. He fails at the Fisher King's castle because he is too shy to ask questions, though his inner self prompts him to ask them. In other words, he lacks the determination and self-confidence of the fully fledged hero. As the hideous maiden on muleback afterwards tells him, he behaved in the Rich Fisher's hall like a man who waits for fine weather and then, when it comes, waits for it to get finer still. One of the factors in his lack of self-assurance is the advice he has been given not to talk too much. Another, apparently, is his guilty anxiety about his mother. The civilizing process to which he has been exposed has robbed him of his original unquestioning confidence in himself and his corresponding carelessness about other people.

In the next stage the pendulum swings too far in the other direction. From being unduly unsure of himself, Perceval reacts to failure and public humiliation by becoming excessively self-reliant. Chrétien's other stories demonstrate his belief in the principle of the mean, of not carrying a good thing too far (and the unknown author of the hermit episode, if he existed, was evidently aware of this). Perceval spends the next five years relying entirely on himself. He never visits Arthur's court, he makes no friends, he sees nothing of Blancheflor. In particular, he puts no trust in religion. He takes his second major step forward when he consults the hermit, discovers his duty to God and the Church and is reminded of his chivalrous obligations to the rest of the human race.

Perceval's story begins as a tale of the Fair Unknown type. The hero does not know his name. His mother calls him Fair Son, and he knows nothing about the rest of his family. His mother is not a fay, but she does live in something close to an enchanted domain, the secluded forest. Once he has reached Arthur's court, however, his adventures branch off on a different track. There is no spellbound heroine transformed into a dragon in a deserted city, no alluring mistress in an otherworld island, and the central character discovers his identity in entirely different circumstances. Guinglain in *Le Bel Inconnu* is told his name at the moment when he has fulfilled his quest. By contrast, Perceval realizes his name at the moment when he discovers that he failed in the Fisher King's hall. Instead of freeing a queen and a city from the grip of death, his action will lay lands waste and cause destruction and sorrow. He discovers who he really is, not in triumph but in defeat. It is through failure that Perceval grows in stature.

Like many leading figures of the Matter of Britain, Perceval grows up in obscurity and comes to his adult field of action from the outside. Not only this, but he is brought up in complete ignorance of chivalry, without the training which a knight was expected to have and with virtually no knowledge of the Christian faith. He is consequently the supreme outsider of all the champions of the Round Table, and indeed he spends little time at Arthur's court. It is often said that his story is a version of the theme of the Great Fool, the hero who is so simple-minded as to be almost insane by worldly standards, but who in his simplicity possesses a wisdom more profound and more potent for good than the wisdom of the world. Whether this is really Perceval's role in the Grail legends is doubtful. The real point seems to be the excitement of watching the best runner win the race from an appalling start.

Like Lancelot and Galahad, his two great rivals for the title of the best knight in the world, Perceval is brought up by a woman alone, without a father. One thing which this suggests is that he is his own man. This impression is reinforced in another account of his adventures, *Perlesvaus*, where he is nicknamed Parluifet, 'made by himself'. That he owes little to his father is shown by his family relationships, which are obviously significant and heavily matrilineal. The Fisher King is his cousin on his mother's side. He is the nephew, sister's son, of the Grail King – the old king who is fed by the Grail – which almost certainly implies that he is the Grail King's eventual heir. The hermit who tells Perceval about the Grail is his maternal uncle. It begins to look as if the mystery of the Grail is a family secret and as though the hero's failure at the Grail castle was, in effect, a failure to claim his inheritance.

The only information of importance we are given about Perceval's father, apart from his premature death, is that he was wounded in battle by a thrust through both thighs. This means that he was crippled in exactly the same way as the Fisher King, but the meaning of this interesting parallel is not explored.

As the story stands, all sorts of other questions are left tantalizingly unanswered. What exactly is the Grail, and in what way precisely is it 'a holy thing'? Why did the lance bleed? What is the connection between the Grail and the lance? Why are there two kings in the castle? Why should asking questions heal the Fisher King? Why should failure to ask them cause destruction and suffering? Who was the hideous maiden on the mule – the Loathly Damsel, as she is usually called – and how did she know what had happened in the Grail castle?

The only approach which seems to make some sense of these questions is the one which finds the origin, though not the full meaning, of the Grail legends in much earlier Celtic traditions. The fact that Perceval comes from Wales is the first of many indications of Welsh influence on the stories. Since Perceval was on his way home when he stumbled across the Grail castle, it may be that the castle was meant to be in Wales. It is possible, though this is highly speculative, that it was a fortress named Dinas Bran in Denbighshire (of which more later).

The castle is an otherworld domain with the Celtic peculiarity of appearing and disappearing. Perceval follows the fisherman's directions to it, but he cannot see it and he is just blaming the fisherman for deceiving him when the castle suddenly appears before his eyes. When he leaves next day and meets his cousin in the wood, she tells him there is nowhere to obtain a comfortable night's lodging within miles. Then he tells her where he spent the night and she realizes that he must have stayed with the Fisher King. Evidently the castle is not always there: or if it is, it is not usually visible to mortal eyes. When Perceval reaches the castle, his host has mysteriously arrived before him, though the hero is on horseback and the Fisher King has to be carried on a litter. Next morning the castle is strangely deserted and the drawbridge is raised by no visible agency. Whatever the meaning of the Grail and the lance, they are not objects in an ordinary castle with an ordinary human lord.

Although the word *graal* was extremely rare, the precise form of the Grail is not described in the story, and this omission was presumably deliberate. It has been argued that the Grail cannot have been a cup because, if it was, the hermit's remark that it did not serve a salmon or a pike to the Grail King would be absurd. The remark implies a large dish, and the writer Hélinand de Froidmont, early in the thirteenth century, defined a *graal* as a wide, fairly flat dish used for serving food at banquets. On the other hand, this definition is untrustworthy because by Hélinand's time the Grail had been positively identified as a chalice. It is not safe to put too much weight on the hermit's comment, which may be a piece of hyperbole. The form of the Grail remains an open question and this is the main point about it: it is a mysterious object.

What is clear is that the Grail is connected with food. It is a serving-vessel of some kind. It carries the Mass-wafer which is the only nourishment of the aged Grail King and it is associated with the fine feast which Perceval and the Fisher King enjoy. It does not serve the feast, as it does in many of the other stories, but it passes before Perceval's eyes with each course of the banquet. The connection with food, and with especially enjoyable and satisfying food, is one of the Grail's most constant characteristics in the stories. It comes out in the episode quoted earlier from Malory, when the holy vessel entered Arthur's hall and 'every knight had such meats and drinks as he best loved in this world'. It links the Grail with the magic vessels of Celtic mythology which provided limitless quantities of delectable food and drink in the otherworld.

Feasting was one of the principal pleasures of the otherworld, and the variety of vessels which provided the otherworld banquet – cauldrons, cups, drinking-horns, platters – probably has much to do with the later variations in the shape of the Grail. Four of them were needed for the wedding feast in 'Culhwch and Olwen': the cup which provided the best and strongest of drink; the plate or table which supplied inexhaustible amounts of food and gave each person the food he liked best; the drinking-horn for pouring the drink; and the cauldron for boiling the meat. Arthur and his men carried off this cauldron. In 'The Spoils of Annwn' they raided the otherworld to seize the magic cauldron which would cook only the food of the brave. The Dagda, 'the good god' of Irish mythology, owned a cauldron which never ran dry and from which no one ever went away unsatisfied.

Just as the food and drink of mortals supports human life, so in mythology the food and drink of the otherworld supports the immortal life and eternal youth of its inhabitants. Zeus and the Greek gods lived on the nectar and ambrosia of immortality. The Norse gods ate the apples of Idun, which kept them for ever young. The Irish god Goibniu brewed beer in a cauldron for the otherworld feast which preserved the Tuatha De Danann, the Irish deities, from ageing and death. The Dagda's inexhaustible cauldron could bring the dead back to life. Another Irish god, Midir, had a similar cauldron, which was stolen from him by the hero Cuchulain. The Welsh god Bran owned a cauldron which reanimated dead warriors who were placed in it.

If the Grail is descended from otherworld vessels of this kind – not from any one of them in particular, but from the idea of them – then it too should be connected with regeneration and eternal life. There is already a hint that it is, when the banquet is served to Perceval and the Fisher King. The table on which the food is placed is made of ivory, resting on trestles of ebony, and we are specifically told that ebony lasts for ever, because it does not rot and cannot burn. There seems to be no point in this comment unless it is meant to suggest that the scene in the castle has something to do with immortality, or 'lasting for ever'. Furthermore, to rot and to burn – to moulder in the grave or to writhe in the flames of hell – are the two alternatives, after earthly death, to eternal life in heaven. We are not specifically reminded that ivory is bone, but it is, and since the bone is the most permanent part of the body, which long outlasts the flesh, this appears to be another hint at immortality.

The Grail is clearly no ordinary object. It is surrounded with mystery, it is holy and it emanates blazing light. Whether the light shines from the Grail itself or from the maiden who carries it is not absolutely clear: *Quant ele fu laiens entree atot le graal qu'ele tint, une si grans clartez i vint* ('When she came in with the grail which she held, so great a radiance appeared . . .').[4] However, it seems to be the Grail which radiates the light, because the procession is immediately compared to the sun (the golden Grail), the moon (the silver carving-dish) and the stars (the candles). This sky symbolism again suggests heaven and eternal life. The Grail is linked with the sun, which from very early times all over the world has been revered as the creator and sustainer of all life on earth. Another old and widespread theme is that the sun's reappearance in the morning after its disappearance at sunset is a symbol of life renewed after death. The early Christians drew a parallel between the sun and Christ, who rose from death and who brought to all mankind the possibility of immortality in heaven.

It is difficult to resist the conclusion that the light of the Grail is the radiance of the Divine Presence, which is also the radiance of immortality. Jesus said: 'I am the light of the world; he who follows me will not walk in darkness, but will have the light of life.'[5] Later, the hermit explains that the Grail carries a Mass-wafer to the Grail

King. The sacred host, on which he has sustained his life long beyond any normal human possibility, is in Christian belief the body of Christ, and the doctrine of the Real Presence of Christ in the bread and wine of the Mass was very much in the air at the time when the *Conte du Graal* was written.

The Grail's link with the Mass is another of its constant characteristics. The pagan otherworld vessels supplied the food of eternal life and happiness. So does the Grail, in providing the Christian's spiritual food, which is God. In this capacity, however, it is a highly unorthodox object. It is associated with lip-smacking feasts as well as with spiritual sustenance. It is borne in solemn procession, but without priests, acolytes, crosses or any of the ecclesiastical accoutrements which might be expected. Odder still, in defiance of all orthodox procedure, it is carried not by a priest but by a woman.

So strange is all this that it has been argued that the scene in the Fisher King's hall represents the teachings of the Cathars and similar heretical sects. These sects were gnostic and dualist, believing in the existence of two great equal powers of good and evil at work in the world. On this view, the radiant Grail is a symbol of light and eternal good. The bleeding lance stands for darkness and eternal evil. The beautiful Grail Bearer is Wisdom, the feminine aspect of the gnostic godhead, and the Grail King is one of the Perfect, the ascetic elite of the Cathars. But the story is not a piece of Cathar propaganda. On the contrary, Perceval is saved from heresy later on, when the hermit instructs him in the doctrines of the Church.[6]

If so, one wonders why the hermit gives no such impression but, quite the reverse, reproaches Perceval for failing to learn the secret of the Grail castle. A more convincing explanation is that Christian authors were adapting to the Christian scheme of things themes which were originally pagan. The results were likely to be unorthodox, especially as the stories were not written by or for theologians, but were meant for a lay audience of nobility and knights. The female Grail Bearer was probably not a Cathar goddess, but in origin a pagan one.

Perceval's family relationships suggest that the scene in the castle has something to do with succession to the kingship. So does the vital question about the Grail. Perceval should have asked, 'Who is served by the Grail?' The answer would have been, in effect, 'The old Grail King, your mother's brother, whose heir you are.' There is a parallel in an Irish story, 'Baile in Scail' (sometimes known as 'The Phantom's Frenzy', though it has nothing to do with phantoms or frenzies), which tells how Conn of the Hundred Battles discovered the Lia Fail, the mysterious stone whose shrieks signified the number of his royal descendants. After this he lost his way in a mist and met a horseman, who invited the king to his otherworld palace. There Conn saw his host, who was the god Lugh, seated on a golden throne astoundingly beautiful. With him was a lovely girl, wearing a golden crown and throned in a crystal chair. She was the Sovereignty of Ireland and she had beside her a silver vat of ale which never ran dry, a golden cup and another golden vessel. She gave Conn a massive helping of beef and pork, and then, filling the cup with ale from the inexhaustible vat, she said, 'To whom shall this cup be given?' Lugh answered, 'Serve it to Conn of the Hundred Battles.' As the girl filled the cup again and again, each time repeating the same question, Lugh named all the future kings who would be descended from Conn. In the end Lugh and the girl and the palace vanished but the magic vessels remained with Conn.

Some form of this story seems to have inspired several features of Perceval's visit to the Grail castle. The hero loses his way and is invited to an otherworld palace by a host who arrives there before him. He sees mysterious vessels of plenty, which a beautiful woman has in her charge. A substantial feast is provided. The crucial question has to be asked, to reveal the hero's identity and establish the rightful succession to a kingdom.

The Sovereignty of Ireland is the prototype of both the Grail Bearer and the Loathly Damsel, the hideous maiden who subsequently reproaches Perceval for failing to ask the crucial question. In *Perlesvaus* we are told that the Grail Bearer and the Loathly Damsel are the same woman. We are not told this in the *Conte du Graal*, but if it had been completed we probably would have been, and their identity explains how the Loathly Damsel knew what had happened in the Fisher King's hall.

There are several Irish tales in which the Sovereignty of Ireland is a beautiful woman who turns herself into a hideous crone. In one of them the hero is Niall of the Nine Hostages, who was a real king of Ireland in the fourth century. As a young man, so the story goes, he went hunting with his four brothers. One of the brothers went to fetch water from a spring. There he saw a scrawny old woman, as black as coal, with foul teeth which spread from one ear to the other, black eyes, a flattened nose, green nails, swollen knees and bulging ankles. The hag demanded a kiss but the young man shrank away from her in horror. The same thing happened to each brother in turn until finally Niall went to get water. He kissed the old crone and made love to her, at which she turned into a radiantly beautiful woman. She told him she was the Sovereignty of Ireland. Her initial ugliness was a sign that the kingship was hard to obtain. Niall had won it by his courage and he, not any of his shrinking brothers, would be High King of Ireland.

The Sovereignty was evidently the goddess Eriu, who personified the land of Ireland as the mate of the High King. Her metamorphosis from ugly hag to beautiful woman represented the change from winter to spring, brought about in the High King's embrace, when the bleak and barren landscape turned green and lovely and rich with life. In the Grail stories the change occurs the other way round. The goddess is first seen as the beautiful Grail Bearer. But the hero fails the test, fails to establish his right to the throne and the Grail, and so the goddess appears to him in her withered and barren guise as a portent of desolation and death.

In the *Conte du Graal* the Grail King, feeble with age, has retired from the world. The Fisher King, who has succeeded him, is crippled. If the Fisher King is not healed, order will break down, lands will be laid waste and many will die. The crucial question, if asked, would have revealed Perceval's true identity, which at this point he himself does not know. It would have shown that he was the rightful heir, so assuring the succession and preventing disorder and desolation.

The question would also have healed the Fisher King. There is a hint here of the old belief that the fertility and prosperity of the land was magically connected with the virility of the king. The Fisher King had been crippled by a spear-thrust in the thighs. It seems clear that the words usually translated 'wounded through his two thighs' were intended to mean 'wounded between his two thighs'. In plain language, he was emasculated. The Fisher King is impotent and his land is threatened with ruin. The two conditions are so closely linked that if one is cured, so is the other.

The puzzling presence of two kings in the castle may be connected with the same belief about a king and his land. The prosperity of the land was threatened if the king grew old and feeble or if he lost his virility. These two states appear to be represented by the Grail King and the Fisher King respectively.

How vague and shifting these traditions were is shown by another early story, in the First Continuation (the first of four sequels to the *Conte du Graal*), in which Gawain goes to the Grail castle. The story begins with Gawain meeting a stranger knight, who was suddenly struck down by a javelin hurled by an invisible enemy. Dying, he implored Gawain to take his place and carry out a vital mission. Gawain agreed, took the stranger's arms and horse, and set off to go wherever the horse might carry him. He rode all night and on the next day through a desolate country, barren and deserted. In the evening he reached the sea. Ahead of him a causeway led far out to sea, with trees growing on either side of it and meeting overhead, making a dark and ominous tunnel. Desperately weary for lack of sleep, Gawain decided to wait until morning before going further, but the horse would brook no delay. It took the bit between its teeth and plunged into the tunnel.

After a long ride, Gawain came to a castle where he was at first joyfully welcomed, but when the people saw that he was not the stranger knight they were disappointed. They left him alone in a spacious hall. In the middle of the room was a massive bier, on which lay a corpse, and on the breast of the corpse was half of a broken sword. Gawain heard loud lamentations and in came a procession of clergy in rich vestments, followed by a crowd of mourners. The priests celebrated the Office of the Dead. After this, tables were set for a meal and a tall king entered, wearing a golden crown. He took his place at table and seated Gawain beside him. Then Gawain saw 'the rich grail', which came into the hall and moved here and there by itself, serving the company an ample meal and filling the cups with wine.

Left alone after dinner, Gawain was awed and frightened by all he had seen. Looking around him, he now observed a lance, standing upright in a silver vessel. Red blood streamed down the shaft into the vessel, and then into pipes which took it away out of the hall. The king reappeared and led Gawain over to the bier, where he asked God to grant that the dead man be avenged, so that the desolate country might be restored to life. The king then produced the other half of the broken sword, which belonged to the stranger knight killed at the beginning of the tale. He asked Gawain to fit the pieces together and mend the sword, but when Gawain tried the two halves did not reunite and the king was grieved. However, he said that if Gawain wanted to know about the strange events he had witnessed, he need only ask.

By this time Gawain was almost dropping with weariness, but he forced his eyes to stay open and asked about the lance and the sword. The king said that no one had ever dared to ask before, but the lance was the one with which Christ was pierced on the cross and it would continue to bleed until the Day of Judgement. The blow of the lance had saved mankind from torment in hell, but the blow which the sword had struck was an evil stroke, which had brought many to their deaths and destroyed the whole realm of Logres. The king was going on to explain this when he saw that Gawain had fallen fast asleep.

When Gawain woke up, it was morning and he was alone with his horse in a field of gorse by the sea. Furious with himself and ashamed, he rode away through the country which had been desolate the evening before but which now was restored to life. The streams were flowing again and the arid woods had turned green. There would have been more people in the country if only he had asked more questions, but those who saw him riding by called out to him: 'Sir, you have both slain us and healed us! Thus you should be glad and joyful for one reason, and sad for the other – glad because of the weal that we now enjoy, for well we know that you are the cause. Yet we should hate you because you did not learn why the grail served. No one could tell the great joy that would have come of asking, but now you must suffer dole and grief.'[7]

There are obviously numerous differences between this adventure at the Grail castle and Perceval's experiences, though both occur in an otherworld of enchantment and mystery. There is no crippled Fisher King, no Grail procession, no Grail Bearer, no old king in retirement. The form of the crucial question is different. There is a mission of revenge which is connected with a broken sword. Whether this sword has anything to do with the one destined to break which the Fisher King gave Perceval is not clear. There seems to be a confusion about the sword and the bleeding lance. In the *Conte du Graal* at one point, Gawain is told that the kingdom of Logres will be destroyed by a mysterious lance. In the First Continuation the blow which laid the kingdom waste was struck by the sword, and yet it was when Gawain asked about the lance that the land came back to life.

It was not until later (in the *Suite du Merlin*) that the logical step was taken of making the lance itself strike the devastating blow. The lance may originally have been the spear of Lugh, which was famed for its atrocious destructiveness. Lugh acquired it by sending three heroes to seize it from its owner, King Pisear. They found it in a room in Pisear's palace, where it stood with its head in a cauldron of water, which was boiling and hissing. The weapon was blazing hot and ferociously destructive because it was the spear of lightning. There are other Irish accounts of what seems to be the same spear, in which it stands in a cauldron of blood, recalling the description in the First Continuation.

If the bleeding lance was originally the lightning spear, however, in the First Continuation it has been transformed into one of the most sacred relics in Christendom, the lance of Longinus, or Holy Lance. This was the spear or lance which a Roman soldier thrust into Christ's side as he hung on the cross, 'and at once there came out blood and water'.[8] According to Christian tradition the soldier's name was Longinus. The lance was miraculously discovered, or so it was believed, at Antioch in 1098, during the First Crusade. The crusaders stormed Antioch but were almost immediately shut up in the city themselves and besieged by a Saracen army. Three weeks went by and both food and hope were running out when a priest claimed to have seen in a vision where the Holy Lance was buried in the church of St Peter. A search was made and the precious relic was found. The discovery restored the crusaders' confidence and emboldened them to sally out and rout the besiegers.

The First Continuation says that the blow of the lance actually killed Jesus (though according to the New Testament he was already dead when the spear pierced his side). The lance is treated as a relic of Christ the Redeemer, whose death on the cross saved all who believed in him from the grip of death and the sway of the Devil. It is a guarantee of eternal life in heaven and its massive regenerating power brings life back to the barren land when Gawain asks about it.

The Grail is even more plainly a magic otherworld vessel than before. It moves about by itself, in a way described as politely brisk, and it provides a lavish feast. Its form is still left vague. It can hardly be a dish, because it fills cups with wine and places them on the tables. It is not a cup, because it serves bread and several courses of food on capacious plates. It remains mysterious.

Another pagan strand now more clearly developed is the theme of the Waste Land, which was hinted at in the *Conte du Graal*. The country has been laid waste by the murder of the dead man on the bier. When he is avenged, it will be restored to life. The theme goes back ultimately to the old Celtic belief that the fertility of the land depended on the vigour and virility of the king. The king was the mate of his land. Down into the twelfth century in one area of Ulster, all the people gathered at the accession of a new king. A white mare was brought and the king went down on all fours and copulated with it. The mare was then slaughtered and cut to pieces. The meat was boiled and the king sat in a bath of the mare's broth, eating the meat and lapping up the broth. The people, standing round, also ate the meat. The ceremony mated the king with his land and ratified his royal authority over it.[9]

The tradition of the king as the mate of his land lies behind the Waste Land theme in the Grail legends, but the theme is incoherent and amorphous. The pattern ought to be this: a king is crippled or ill; as a result his land is barren; the hero heals the king and fertility is restored to the land; probably, the hero's feat shows that he is the rightful heir. There is no Grail story in which this simple and satisfactory pattern appears (nor has any Celtic story survived which contains it). In the First Continuation there is a waste land which is restored, but no crippled or ill king and consequently no healing. In *Parzival* there is a crippled king who is healed by the hero, but there is no waste land. In *Perlesvaus* there is an ill king and a waste land, but no healing. It was not until comparatively late in the development of the Grail legends (in the *Suite du Merlin*) that an author took the ingredients of the pattern from his predecessors and put them together in something approaching a satisfactory shape.

The legends of the Grail, then, were constructed initially from loose and shifting pagan traditions to do with immortality, fertility and kingship. There is the idea of a magic vessel which is the source of eternal life. Sometimes it is in the charge of a fertility goddess, who turns into the Grail Bearer, sometimes it is not. It is guarded in the castle of a king, who may or may not be impotent or ill and whose land may or may not be desolate or threatened with ruin. The healing of the king is sometimes, but not always, magically connected with the restoration of life to the land or the removal of the threat. The successful hero is the king's rightful heir, who alone can keep the land safe and prosperous. His reward is to inherit the Grail and with it immortality.

The pagan themes were not obliterated but transformed by Christian ideas. The Grail as the ultimate source of life is linked with the Divine Presence, the Mass and immortal life in heaven. It is still the source of the most delectable physical food, but also of the most sublime spiritual food. The fearsomely destructive spear becomes the Holy Lance which struck the blow that killed Christ on the cross. It is consequently the agent of Christ's redeeming sacrifice, which reprieved humanity from a sentence of death, and its life-creating power extends to the physical plane, for it gives life to a barren land. The motif of the crucial question, the question that must be asked to reveal some great secret and bring some great benefit to men, is close to the heart of Jesus's teaching. 'Ask, and it will be given you; seek, and you will find; knock, and it will be opened to you.'[10]

God can be ours for the asking, but we must ask. Human nature being what it is, asking will be hard. Perceval is too unsure of himself to ask, and Gawain is too exhausted. There is a significant scene in *Perlesvaus*, when Gawain fights his way valiantly to the Grail castle. He is welcomed and taken to the hall, where he sits down to feast with twelve knights of the castle. Two maidens bear in the Grail and the bleeding lance. When Gawain sees the Grail, he feels such rapture that he can think of nothing but God. The twelve knights wait anxiously for him to ask the vital question, but he is lost in his reverie and he says nothing. The knights go sorrowfully away. When Gawain comes to himself, he looks around him and sees a chessboard whose pieces play by themselves. He plays against them and they checkmate him twice, and the third time, seeing that they are winning again, he sweeps them angrily off the board. In the morning he is told to leave the castle and, ashamed, he rides away.

Gawain's failure seems far from unpardonable, but it is a failure. The sight of the Grail so wraps him in thoughts of God that he forgets the real purpose of his quest, which is not to think about God but to ask the question. He does not ask it, though he has previously been told exactly what to say.

The deepest link between the Grail stories and the rest of the Matter of Britain is that the quest of the Grail is the supreme example of a search for the true, ideal self, which leads to a triumph over death. Perceval's progress in the *Conte du Graal* is towards becoming the best knight in the world. When Gawain fails to mend the broken sword in the First Continuation he is told that his prowess is not yet great enough, but that he may succeed in the future. He restores the desolate country to life by asking about the lance, but if only he had asked about the Grail, something still more wonderful would have happened: 'no one could tell the great joy which would have come of asking'. The 'great joy' recalls the joy of the people of the Waste City, restored to life by Guinglain, and the delirious rejoicing over Erec's success in the adventure of the Joy of the Court. Both these exploits were victories over death.

If Gawain had succeeded, he would have conferred an inestimable blessing on the people, not only on himself. They would have shared the fruits of his triumph. They tell him that, as it is, he has healed them but also slain them. If he had asked the crucial question, so it seems, they would have obtained immortal life.

If Gawain had achieved this, he would have been a second Christ or a greater than Christ. Later, in the *Queste del Saint Graal*, the Grail hero is virtually a reincarnation of Christ on earth. The Grail legends were unorthodox, not because they drew on pagan themes, but because they implied a pagan view of the hero as godlike and in their scheme of salvation they left no room for the Church. The hero who wins the Grail does so partly by right of birth, because he is the heir of the Grail King, and partly by courage, resolution and strength of character, because he is the best knight in the world. He does not achieve it through the channels of the Church, which plays little or no part in the story.

It is typical of the legends that in the *Conte du Graal*, when the time comes for Perceval to be converted to Christianity, he is not instructed in the faith by a priest, an authorized official of the Church. He is taught by a hermit, an unauthorized maverick, who has retreated from both the world and the Church to find his own path to God in solitude. The hermit is the religious counterpart of the knight errant, the man who goes out into the wilds to work out his own salvation.

The message of the Grail stories is that the hero can reach the ultimate source of physical and spiritual life by his own efforts and without the aid of the Church. This was not a view of spiritual progress which the Church was anxious to encourage.

Joseph of Arimathea

So far we have not been told anything about the history of the Grail – where it came from and what gave it such holiness and power. About 1200, however, Robert de Boron wrote a long poem, *Joseph d'Arimathie*, in which he told how the Grail came into the possession of Joseph of Arimathea, the first Keeper of the sacred vessel. Apparently Robert was not the first person to connect the Grail with Joseph, for he said he had seen a 'great book', written by 'great clerks', which contained 'the great secret that is called the Grail'.[11]

Joseph of Arimathea is a minor character in the New Testament, a wealthy Jew who was a covert admirer of Jesus. With Pilate's permission, he took the body of Jesus down from the cross. Joseph and Nicodemus, another of Jesus's secret disciples, buried the body in a new tomb, which Joseph had intended for himself and from which it subsequently disappeared. This is all the New Testament says of him, but by the fourth century there was a story that the Jews were furiously angry with Joseph and locked him up in a windowless room, threatening to kill him. While Joseph was praying in the darkness, he saw a flash of light and someone took him by the hand. It was Jesus, risen from death, and he led Joseph back to his own house. When the vengeful Jews came to fetch Joseph, the locked room was empty.[12]

Robert de Boron's story is much more elaborate. He says that the 'vessel' in which Jesus celebrated Mass at the Last Supper was afterwards given to Pilate by a Jew. Pilate gave it to Joseph of Arimathea when Joseph asked permission to take Jesus's body from the cross. While Joseph and Nicodemus were preparing the body for burial the wounds began to bleed and Joseph collected the blood in the vessel. He then put the body reverently in the tomb and hid the vessel in his house.

When Joseph was in prison, the risen Christ came to him, carrying the 'great and precious vessel', from which blazed a radiance of light that lit up the whole cell. The Saviour gave Joseph the vessel, saying that it was to be called a chalice and that, in token of the Trinity, only three men were ever to have the guardianship of it. Everyone who saw it would be in Christ's presence and would have lasting joy and fulfilment. The Saviour also taught Joseph secret words, which are not revealed in the poem but are described as 'holy words ... which are tender and precious, gracious and merciful, which are rightly called the secret of the Grail'.[13]

Joseph remained in prison for many years, but because he had the chalice with him he needed nothing to eat or drink. Eventually, he was freed by the Romans. Joseph had a sister, Enygeus, who was married to a man named Bron (the name is given less frequently as Hebron). The three of them, with other Jewish converts to Christianity, left Palestine and settled elsewhere – we are not told where. Unfortunately, some of the company were guilty of the sin of lust and as a result the crops which the community planted would not grow and they were all in danger of starving. In this emergency Joseph was inspired by the Holy Spirit to set up a table for the service of the holy chalice, in commemoration of the Last Supper. On the table were placed the chalice and a fish, which was specially caught for the occasion by Bron, who was therefore called the Rich Fisher. One place at the table was left empty, to represent the seat of Judas Iscariot, the traitor.

When everything was ready, Joseph called his followers to the table. Those who believed in the Trinity and were free of lust took their places and were filled with delight, with a rapturous happiness like the joy of a fish which is caught on the hook but escapes back into the water. The rest, however, could not approach the table. They could not see the chalice, apparently, and they were filled with shame. From then on, those who were chosen attended the table every day and evidently received both spiritual and physical nourishment. The vessel was now called the Grail, because it delighted those who saw it. This is a clumsy pun on the words *graal* and *agréer*, 'to delight'.

One place at the table was still empty and a man named Moses, who had failed the test, begged to be allowed to take it. When he sat down, the earth opened and swallowed him up. A voice from heaven announced that only the grandson of Bron and Enygeus would be worthy to occupy the empty seat.

Years went by. Bron and Enygeus had twelve sons. Eleven of them married, but the youngest, Alain, had a pious horror of sex. He said he would rather be flayed alive than marry. The divine voice, however, declared that he would have a son, who would be the Keeper of the Grail. Alain, leading his brothers and their wives, set out for the far West, to preach the word of Christ. A man named Peter also left, to go to the vales of Avaron in a wild country in the West, where he was to await the coming of Alain's son. Finally, Bron, the Rich Fisher, was entrusted with the Grail and taught the secret words which Christ had revealed to Joseph years before. He too went to the West, with the Grail. There he was to wait for his grandson, to whom he would hand on the holy vessel. Joseph of Arimathea was left behind.

There are many obscurities and confusions in this story, but it does provide an account of what the Grail was and how it passed from Palestine to 'the far West', which almost certainly means Britain. The Grail is still a profoundly mysterious object, 'a great secret', whose inner meaning is enshrined in 'secret words' known only to the Grail Keepers. However, it is now explicitly a Christian relic. It is a vehicle of the Divine Presence and so it radiates blazing light. It is doubly linked with Christ's blood, shed on the cross to save men from death. The blood which oozed from the dead Saviour's wounds was caught in it and it was also the cup of the Last Supper, from which was poured the wine of the first Mass which was not merely symbolically but in all reality the blood of the Redeemer. The worshipper who drinks the blood becomes one with Christ. The Grail is therefore the vessel of union with God. The supreme gift which it confers is immortality, eternal life in heaven, as is shown by the metaphor of the fish, which is caught on the sharp hook of death but joyfully escapes to live again.

The Church is once more left out in the cold. The 'great and precious vessel', the holiest of relics and most powerful of Christian talismans, is guarded not by the Church but by a line of independent Grail Keepers, who trace their authority direct from the risen Christ. Access to the Grail is not through the Church. The secret words which the Keepers alone know (and which recall the prayer full of secret divine names taught to Perceval by the hermit) are not the property of the Church. There are to be only three Grail Keepers, which implies that with the coming of the third some great consummation will take place. We are not told what this will be, but it does not appear to have anything to do with the Church.

The story gives the first clear-cut indication of a belief which swiftly became important in the legends, that sexual purity is an essential qualification for beholding the Grail. The sin which bars a believer from the Grail's presence, and so from eternal life, is lust.

Although the Grail is a Christian relic, it remains closely linked with pagan themes. Its guardianship is still a family secret. It has a mysterious power of discrimination, of separating the sheep from the goats. Some of the otherworld vessels of Celtic tradition had the same quality. The cauldron of Annwn would not boil the food of a coward. The Irish god Manannan mac Lir owned a golden cup which could distinguish truth from lies. Like the pagan vessels, again, the Grail is associated with immortality, with a state of sublime happiness and with a supernatural supply of food and drink. The Grail kept Joseph alive for years in prison and at the table which he set up it transported the faithful out of the everyday human world into a divine world of eternal life and bliss.

The episode of the earth engulfing the sinner who rashly took the empty seat at the table obviously echoes the story of Korah, Dathan and Abiram in the Old Testament. They were opponents of Moses, and God silenced them by making the earth open and swallow them up.[14] The fact that the name Moses has now been transferred to the offender who is punished in this way suggests that he was meant to stand for the Jews, whose special relationship with God in the Old Testament had in Christian eyes been cancelled by the coming of Jesus and his rejection by the Jews. The motif of a special seat intended for a hero, however, may have come from Celtic mythology. There was an Irish story of how the god Lugh, when he first came to the hall of the Tuatha De Danann, sat down in the Seat of the Sage, which was reserved for the wisest. The assembled gods recognized his right to the seat and hailed him as the champion who would lead them against their enemies.

In the episode of the fish caught by Bron, pagan material was again adapted to a Christian purpose. The fish is an old symbol of Christ, and the fish which Bron caught is the divine body, the bread of the Mass. It is placed on the table with the Grail itself, in which is the divine blood, the wine of the Mass. The rich Fisher King, whose hall contained a wealth of precious objects, is now called the Rich Fisher, 'rich' in a spiritual sense.

The name Bron probably comes from a figure of Welsh tradition, Bran son of Llyr. The alternative form Hebron was presumably meant to give the name a biblical ring. Bran was originally a Celtic god, but by the time he appeared in 'Branwen', one of the stories in *The Mabinogion*, he had been turned into a king of Britain and for some unknown reason had been given the Christian title *Bendigeid* ('Blessed'). Bran was of gigantic size, so huge that no house was big enough to contain him. He held court at Harlech in Wales and also at Caer Seint, or Segontium, the original of the Waste City in the Fair Unknown story. He owned a magic cauldron which restored the dead to life: 'a man of thine slain today, cast him into the cauldron, and by tomorrow he will be as well as he was at the best, save only that he will not have the power of speech'.[15] This cauldron of regeneration is one of the ancestors of the Grail, and its inability to restore the power of speech may have something to do with the failure of the Grail hero to speak at the critical moment.

Bran had a sister, Branwen, the most beautiful maiden in the world. He gave her and the cauldron to the King of Ireland. After a time the Irish king began to treat Branwen cruelly. When Bran heard of this, he gathered an army in Wales and invaded Ireland. He himself waded across the Irish Sea, looking like a mountain on the move. In a great battle in Ireland the magic cauldron was smashed to pieces. Bran was wounded in the foot and, as a direct result apparently, the land of Britain fell waste. Branwen was rescued, but almost all the warriors on both sides were killed in the battle. There were only seven survivors of Bran's army. One of them was his nephew Pryderi, who is probably the original of Perceval.

Bran ordered the survivors to cut off his head. They did so, and took the head back to Wales, where Branwen died of grief for the carnage which had been wrought on her account. The seven took Bran's head to Harlech, where they lived for seven years. Birds sang beautiful songs to them and they feasted merrily on lavish supplies of meat and drink with the head, which was just as good company as it had been when Bran was alive. At the end of the seventh year they took the head to Grassholm, an island off the Pembrokeshire coast. There they found a royal hall overlooking the sea. It had two doors which were open and a third, on the side towards Cornwall, which was shut and which they knew they must not open. They spent eighty happy years there in the company of the head. They were not conscious of the passing of time and they did not grow older. This was called the Feasting of the Wondrous Head. Then one of them opened the forbidden door and suddenly they all remembered the tragedy they had experienced and the kinsmen and friends they had lost.

The seven left Grassholm and took the severed head to London. They buried it in the White Hill (presumably Tower Hill) with its face towards France, and as long as it remained there no plague could cross the sea to Britain.

Here we have a story about a king who had superhuman attributes, owned a magic life-giving vessel and was wounded with a poisoned spear. When he was wounded his land became barren, which suggests that the wound in the foot is a euphemism for emasculation. Bran suffered the same injury as the Fisher King. His severed head remained mysteriously alive and presided over an otherworld feast, at which his followers were supernaturally regaled with food and drink in a state of rapturous happiness, knowing no grief or care and growing no older.

It looks as if Bran was the prototype of the Fisher King or Rich Fisher, and his transition from pagan hero to custodian of the Grail would have been eased by the fact that he was already known as Bran the Blessed. The title Fisher King, however, is mysterious. The *Conte du Graal* accounts for it by saying that fishing was the only sport which the crippled king could enjoy, but this explanation seems as lame as the king himself. It has been suggested that the title involved a medieval pun on *pescheor*, 'fisher', and *pecheor*, 'sinner'. The king represents sinful humanity at large and he is wounded in the genitals, which are the fount and origin of sin. Wolfram von Eschenbach seems to have recognized the pun, for his Maimed King was pierced through the testicles with a poisoned spear as a punishment for lechery. Unfortunately, this theory does not explain why the title Fisher King was chosen instead of Sinner King in the first place.

Another suggestion is that the fish was a symbol of Christ and that the king acquired his title because medieval writers connected the crippled ruler with the wounded Christ on the cross. Hence he had the Holy Lance in his keeping as well as the Grail, both of them relics of the Passion. There are difficulties about this explanation too. It is not easy to construct a convincing parallel between Christ and the Fisher King. Christ was not emasculated or wounded in the genitals or the thighs. If the Fisher King is Christ, then his father, the old Grail King, must be God the Father, but it is hard to see why the

Almighty should be feebly confined to one room in his Son's castle, sustained on the sacred host.

A third explanation is that the fish is an age-old symbol of life and fertility, either because of its multitudes of eggs or because of its phallic shape and sperm-like penetrative motion through the water. The problem here is that although fish symbolism of this kind is known from other places and periods, there seems to be no evidence of it in Celtic mythology or in medieval Europe.

A fourth explanation traces the title back to Bran himself. He was the son of Llyr, which means 'son of the sea', and this link with the sea could have given him a connection with fishing. There is a ruined castle near Llangollen in North Wales called Dinas Bran, the fortress of Bran. The castle stands on top of a hill above the River Dee, which is noted for its fishing. A tradition might have grown up that Bran enjoyed angling on the Dee and, if so, this may lie behind the scene in the *Conte du Graal* of the crippled king fishing from a boat in a broad river. All this is extremely hypothetical, but the failure of medieval authors to produce an adequate explanation of the Fisher King's title suggests that they did not invent it and that it came from the pagan past: perhaps in some connection with Bran which we would understand if more traditions about him had survived.

The link between Bran and the Fisher King inspired R. S. Loomis to argue that the Grail legends were founded on a philological blunder. He pointed out that a late list of the Treasures of Britain includes the drinking-horn of Bran, which supplied whatever food and drink anyone desired. In Old French the words for 'horn' and 'body' were the same, *cors*. French story-tellers, Loomis thought, muddled the two and mistakenly turned the magic drinking-horn (*cors*) which provided the otherworld feast into the body (*cors*) of Christ, the Mass-wafer which keeps the old Grail King alive in the *Conte du Graal*. Joseph of Arimathea was brought into the story because he was known from the New Testament to have had temporary custody of Christ's body, which he took down from the cross.[16]

This theory seems too ingenious to be true. It ascribes to the French story-tellers a degree of muddle-headed stupidity which is not easy to credit. Bran's drinking-horn is not mentioned until the fifteenth century, long after the Grail legends had evolved, and how much further it goes back is not known. More to the point, if the Christian concept of the Grail was based on a confusion between the magic horn and the body of Christ, it is extremely difficult to account for Robert de Boron's story. The earliest surviving explanation of the Grail, and of Joseph of Arimathea's connection with it, does not link the Grail with the divine *body*, but with the divine *blood*, by identifying it as the chalice of the Last Supper. The variations in the form of the Grail in the legends suggest that it was not descended from any one other-world vessel but from a variety of them. It was therefore possible for different writers to picture the Grail in different ways or to leave its form deliberately obscure.

The identification of the Grail as a cup may have been inspired by one of the strangest features of the story in the *Conte du Graal*, where the holy vessel of the Divine Presence is carried in procession by a woman who was originally a fertility goddess. A cup is an obvious symbol of woman and of the channel through which new life enters the world, well known to Christian writers in this light through the Book of Revelation.[17] In both its pagan and its Christian contexts the Grail is linked with regeneration and immortality. The cup is an emblem of life and fertility which could be given an exalted spiritual meaning as the cup of the Last Supper.

Perceval and the Grail

Robert de Boron followed his account of the Grail's origins in the *Joseph* with another poem, *Merlin*, in which the Round Table is instituted on the model of the Grail table, with one seat again left empty. This is the Siege Perilous, reserved for the hero who is to win the Grail. The identity of this hero is not revealed, but it seems quite likely that the part was intended for Perceval. The motif of the Siege Perilous shows that only one hero can win the Grail and implies that he not come from the established circle of Arthur's knights, who already have their seats at the Round Table. This fits neatly in with the tradition of Perceval as the supreme outsider.

Perceval is the Grail hero, the destined third Keeper, in a sequel by an unknown writer, the Didot *Perceval*. Here Perceval is the son of Alain le Gros and the grandson of Bron, the Rich Fisher King. Bron is not crippled but he is now feeble and debilitated, suffering from extreme old age but unable to die until he can entrust the Grail to his successor. There is also a significant change of detail about the Siege Perilous. Instead of corresponding to the seat of Judas Iscariot at the Last Supper, it now represents the place of Jesus himself, which no one has yet been found worthy to occupy. The implication is that the Grail hero will be virtually a second Christ. (Shrinking away from this perhaps, the author later makes the Siege Perilous the seat of Judas again.)

After the death of his father, according to this story, Perceval went to Arthur's court, where he quickly distinguished himself and was made a knight of the Round Table. He then asked Arthur's permission to sit in the Siege Perilous. Against the king's advice but with the approval of Lancelot and Gawain, who admired him, Perceval crossed himself and sat down in the empty chair. Immediately the stone seat split in two with a cry of agony that terrified all who heard it. A huge shadowy cloud emanated from the ground and a mysterious voice declared that the enchantments of Britain would not end until the best knight in the world came to the house of the Fisher King and asked: 'What is done with the Grail and whom does it serve?' The enchantments of Britain are not explained, but apparently Perceval's action had cast the land under a spell, and this is a variation on the Waste Land theme.

Leaving Perceval's subsequent adventures aside for the moment, this episode recalls the legend of the Lia Fail, or Stone of Destiny, which cried out when the rightful king took his place on it: for Perceval was the heir of the Fisher King. It is also one of the indications that the prototype of Perceval was the Welsh hero Pryderi.

The first four stories in *The Mabinogion* contain almost all that is left of the saga of Pryderi. He came from Dyfed, or south-west Wales, which was the part of Wales most heavily influenced by the Irish, and there are resemblances between him and the Irish hero Cuchulain. Pryderi's mother was the beautiful and fascinating Rhiannon, who was originally a Celtic mother goddess, closely associated with horses. Who Pryderi's father was is not clear. Ostensibly he was the lord of Dyfed, and Pryderi inherited Dyfed from him, but there are hints that the hero's real father was the King of Annwn, the ruler of the otherworld. This would give him a claim to the otherworld kingdom and its magic cauldron, just as Perceval is later the Grail Keeper's heir. In *The Mabinogion* Annwn is a fine palace, richly equipped, near the eastern border of modern Pembrokeshire. Apparently it could not always be seen, but might appear suddenly to mortal eyes, like the Fisher King's castle.

Later on, Pryderi's mother Rhiannon married Manawydan son of Llyr (the Welsh counterpart of Manannan mac Lir, the Irish god of the sea). Manawydan was the brother of Bran the Blessed, who was the prototype of the Fisher King. Pryderi's family relationships are now complicated. He is the son of an otherworld king who owned a magic cauldron. He is also the nephew of Bran, who owned a magic cauldron and presided at an otherworld feast, as Perceval in several of the later stories is the nephew of the Fisher King or the Grail King. This tangle of traditions may lie behind the curious detail in the *Conte du Graal* that Perceval's father was wounded in the same way as the Fisher King. It seems that the two had partly coalesced.

Pryderi and Manawydan accompanied Bran on the expedition to Ireland and were two of the seven survivors who took Bran's severed head back to Wales and feasted for years in its company. It was after this that Pryderi bestowed his mother Rhiannon on Manawydan, and with her authority over Dyfed, which was a fertile and prosperous country. After feasting at their court at Arberth (modern Narberth) Pryderi and Manawydan and their wives went out on to the Mound of Arberth, which had a peculiar property. If a man of royal blood sat on it, he either received blows and wounds or he saw a wonder. In effect, it was a kind of Siege Perilous.

As soon as Pryderi and the others sat down on the mound, there was a peal of thunder and they were enveloped in a thick mist. When the mist cleared, the whole land of Dyfed had fallen desolate. There were no flocks or herds to be seen, no people, and no houses except the empty buildings of the court below.

The parallels between this story and the episode in the Didot *Perceval* are obvious. In both cases the hero sits in a perilous place. An alarming sound is heard – thunder or a cry – a mist or cloud gathers, and the land is cast under an enchantment. In the Welsh tale it turns into a waste land.

The four survivors lived on wild beasts, fish and wild honey. One day Pryderi and Manawydan were hunting a shining white boar. They pursued it to a massive fortress, which had appeared in a place where they had never seen a building before. Against Manawydan's advice, Pryderi went into the fortress. Inside he saw a fountain and a beautiful golden bowl, which hung by four chains whose upper ends he could not see. He took hold of the bowl, and at once his hands stuck to it, his feet stuck fast to the floor, and he lost the power of speech.

Manawydan returned home and told Rhiannon that Pryderi had not come out of the fortress. She reproached him bitterly for leaving his stepson in the lurch. Hurrying to the fortress, she went in and saw Pryderi, motionless and silent. She grasped the bowl and she too was held fast. Night came on, there was a peal of thunder and a fall of mist, and the fortress vanished with Pryderi and Rhiannon still in it. The two of them were prisoners in the otherworld. In the end, however, Manawydan broke the enchantment. Pryderi and Rhiannon were returned to him and Dyfed was restored to life, with its people and houses, flocks and herds, all as they had been at their most prosperous.

Bits and pieces of this story seem to have influenced the Grail legends. Again there is an otherworld castle which suddenly appears and disappears. In it is a magic vessel which renders the hero speechless, recalling Perceval's inability to ask the vital question in the presence of the Grail. Eventually the spell is broken, though not by the hero himself in this case, and the waste land comes back to life. Rhiannon's anxiety over Pryderi may perhaps have contributed to the later story about the grief of Perceval's mother when he left home.

Returning to the Didot *Perceval*, after the hero sat in the Siege Perilous and the mysterious voice was heard, all the knights of the Round Table vowed to find the house of the Fisher King and ask the question which would end the enchantments of Britain. The next part of the story roughly follows the tale of Perceval's adventures in the *Conte du Graal*. After various exploits Perceval came to a river where he saw Bron, the Fisher King, in a boat. He was invited to Bron's castle for the night and a feast was served. As the first course was brought in, a girl entered the hall carrying two silver carving-dishes. She was followed by a young man bearing the Holy Lance, which dripped blood, and then by another young man holding on high the Grail itself, the vessel containing the divine blood which the risen Christ had given to Joseph of Arimathea in prison. Despite hints from his host, Perceval did not ask the crucial question, partly because his mother had warned him not to ask questions and partly because he was desperately sleepy. The next day he was reproached by a maiden, who told him that if he had asked, the Fisher King would have been restored to vigorous health and the enchantments of Britain would have ceased.

For years Perceval tried to find the Grail castle again, but in vain.

At last, following Merlin's directions, he came to the Fisher King's castle for the second time. He saw the procession of the Grail and asked the question. The Fisher King was instantly cured of his debility, and we are told that he was as spry as a fish. There was great rejoicing in the castle and Perceval identified himself as Bron's grandson. Bron taught him the secret words which enshrined the inner mystery of the Grail. He entrusted the precious vessel to Perceval's keeping and from it there came a melody and a scent so sweet that it was as if they had been in Paradise. Now at last Bron could die in peace, and Perceval saw angels bearing his soul up to heaven, where he would enjoy eternal life and happiness. Meanwhile, at Arthur's court the split in the Siege Perilous closed with a resounding crash, and Merlin told Arthur that the Grail quest had been achieved and the enchantments of Britain were at an end.

In this version of Perceval's quest the Grail is borne by a young man instead of a woman. The crucial question is asked, but it is not answered: or rather, we are not told the answer, which is presumably contained in the secret words taught to Perceval by Bron. The point of asking the question is not to save the Fisher King's life, for he dies very soon afterwards, but it does end the enchantments of Britain, which is analogous to restoring a waste land to life. It also enables Perceval to be recognized as Bron's successor and Bron to die and receive his reward of immortality.

Several other authors told the story of Perceval's quest, on similar lines but with considerable differences of detail. These include variations in the form of the Grail. In *Perlesvaus* the Grail is the vessel in which Joseph of Arimathea collected the blood of the crucified Christ, which suggests a cup. Joseph kept it and also the lance which pierced the Saviour's side. In the procession in the Fisher King's hall the Holy Lance and the Grail are carried by two maidens, side by side, with the point of the lance bleeding into the Grail. Here the two relics are connected in a logical way, with the lance which drew blood from Christ's side bleeding into the cup in which the blood was collected. However, when Arthur goes to the Grail castle and sees the sacred vessel, it appears to him in five different shapes. What the first four are is not revealed, but the last and most significant of them is as a chalice. The Grail is evidently still a profoundly mysterious object which can take various forms.

In Wolfram von Eschenbach's *Parzival* the Grail is not described

Perceval's Family Tree

According to Robert de Boron and the Didot Perceval

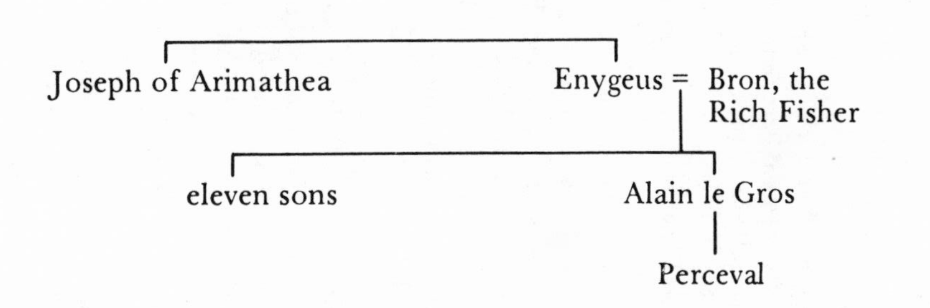

According to Perlesvaus

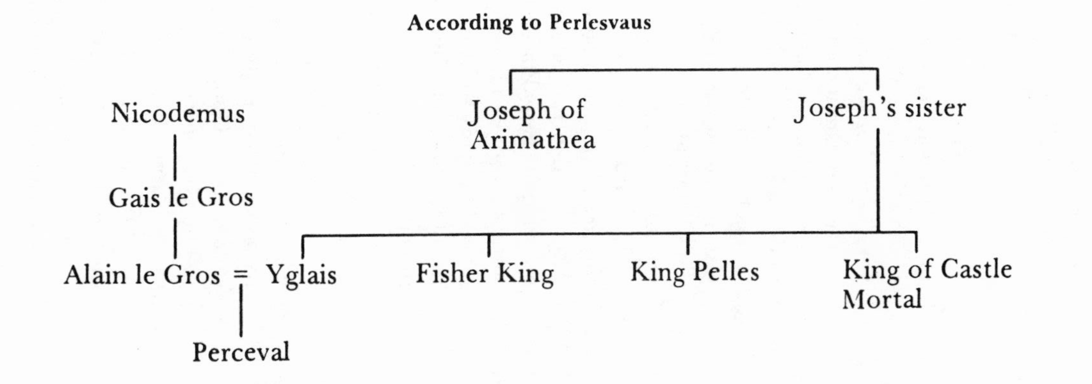

when Parzival (Perceval) first sees it in the Maimed King's castle. It is 'a thing called the Grail'. It is 'the perfection of Paradise' and it 'surpasses all earthly perfection'. It is carried by a beautiful woman, the Maimed King's sister, whose office requires her to be chaste. Its appearance is followed by a magnificent feast, at which everyone is supplied by the power of the Grail with whatever food and drink he most desires at that moment: 'for the Grail was the fruit of blessedness, such abundance of the sweetness of the world that its delights were very like what we are told of the kingdom of heaven.'[18]

The old connection between the otherworld vessel, sumptuous feasting and immortality is so clearly expressed here that it is surprising to discover later that the Grail is not a cup or a dish but a stone. It was brought down to earth by the angels who were neutral in the war in heaven, when the Devil and his legions rebelled against God. This detail about the stone from the sky recalls ancient and widespread beliefs about the divine and magical powers of meteorites.

The Grail is guarded at the castle of Munsalvaesche (from *mons silvaticus*, 'wooded mountain') by chaste knights called Templeisen. This is evidently a reference to the Order of Knights Templar, who guarded the Holy Places in Palestine and were sworn to celibacy. The stone's power sustains all who live in the castle and enables the immortal phoenix to be reborn from its own ashes. It restores youth to the old. Those who live close to it do not age, except that their hair may turn grey, and they cannot die. The emasculated Maimed King lives on in appalling agony because of the Grail, his anguish slightly eased when the point of the bleeding lance is thrust into his wound. The Grail provides all food and drink on the earth and all creatures which are good for food, whether they fly or run or swim. It draws its power from a Mass-wafer, which is brought down from heaven each year on Good Friday and placed on the stone by a dove.

The Grail here is the vivifying energy which sustains all life on earth. Nature and the animal kingdom owe their teeming abundance to it. It is a talisman of regeneration and immortality, linked with the bread of the Mass, the body of Christ and the life-bringing sacrifice of the Saviour on the cross. It is called *lapsit exilis*, a meaningless phrase which recalls an episode in the legend of Alexander the Great. The *lapis exilis* or 'paltry stone' was sent to Alexander from the Earthly Paradise. It resembled a human eye. When weighed, it was heavier than any amount of gold, but if a little dust was sprinkled on it, it was lighter than a feather. The moral is that the eye, like Alexander himself, is greedily insatiable and tries to encompass the whole earth, but a handful of dust, dropped on it in death, blinds it and blots it out. The stone taught Alexander humility, and humility is one of the virtues which Perceval has to acquire in his quest.

Though this episode in the legend of Alexander probably influenced the story, the Grail in *Parzival* is not remotely adequately explained as a symbol of humility. It is much closer to the Philosopher's Stone of alchemy, which is both a mysterious object of gigantic power and a spiritual state. The Philosopher's Stone was believed to turn anything it touched into gold, to cure all diseases and to confer eternal life and youth on its possessor. It also stood for a 'golden' spiritual condition, the highest and most perfect conceivable, which was the state of union with God or of virtually being God. Like the Grail hero, the alchemist trod his own path to salvation, independently of the Church, and was consequently suspect. Alchemy first attracted attention in western Europe in the twelfth century and *Parzival* was written early in the thirteenth. The phrase *lapis exilis* was later used for the Philosopher's Stone and 'the paltry stone' could be connected with Christ as 'the stone which the builders rejected' that became 'the head of the corner'.[19] It seems likely that alchemical lore influenced the concept of the Grail as a mysterious stone which gave life, physical and spiritual, which was closely associated with Christ, and which was guarded by a community of the wise, outside the Church's sway.

The Grail's connection with the Earthly Paradise is again brought out in *Perlesvaus* and seems a natural development from the holy vessel's original location in the Celtic otherworld, where life was lived free of death and the other evils of the human condition. The Earthly Paradise, the garden of Eden in all its joyous perfection, where there was no death, no pain and no work, was believed still to exist somewhere on the earth, just beyond the borders of the known world. In *Perlesvaus*, too, the Grail's attendants do not age. Twelve knights live in the Grail castle, feasting in the sacred relic's presence. They are grey-haired and they look about forty years old, but each of them is a hundred or more. This episode brings to mind the Welsh story of the seven companions of Bran, the prototype of the Fisher King, feasting in the presence of his head and growing no older.

In both *Parzival* and *Perlesvaus* the hero is the nephew, sister's son, of the Maimed King or Fisher King, and he has a hereditary right to the Grail kingship. There is no waste land in *Parzival* and the hero's task is to heal the Maimed King and establish his right to the throne. There is a significant change in the question which he must ask his crippled uncle. It is not 'Whom does the Grail serve?' but 'Sir, why is it you suffer so?'[20] The point of the change is that the second great virtue which the hero must learn, beside humility, is compassion. When he has learned both he returns to the Grail castle and asks the question. The Maimed King is immediately healed and Perceval is proclaimed King of the Grail in his stead.

Things are very different in *Perlesvaus*, where the Waste Land theme appears in a form close to 'the enchantments of Britain'. Before the story opens, the hero (here called Perlesvaus) has already visited the Grail castle and failed to ask whom the Grail served. As a result all sorts of evils have occurred. The Fisher King is stricken with disease and Perceval himself has fallen ill. King Arthur has lapsed into a state of apathetic sloth and most of his knights have deserted him. The land is plunged in war and sorrow. The progress of Christianity, whose triumph over Jews, Muslims and pagans is the principal duty of Arthur and his champions, has been gravely impeded. The theme of the question revealing the rightful heir still persists. The root of the trouble is that because Perceval did not ask the question, the Fisher King did not realize who he was.

Perceval does not win the Grail by returning to the castle and putting the right question. The question is never asked and the hero achieves the Grail by a mixture of magic and force of arms. The Fisher King dies and the Grail castle is captured by his evil brother, the King of Castle Mortal. The 'castle mortal' suggests the body, the flesh, and its ruler is apparently the lord of sin and death. The castle of the Grail has now been seized by a usurper and the Grail itself disappears. Protesting that he is the rightful heir, Perceval visits another of his maternal uncles, King Pelles the hermit, who provides him with a white mule and a banner. Equipped with these and helped by a righteous white lion, Perceval storms the Grail castle single-handed. The evil king commits suicide. Perceval makes himself master of the castle and, now that the true heir is installed, the Grail reappears.

Perceval's success does not instantly banish the evils from the land, however. Arthur and Gawain visit him and see the Grail in its five different forms. After this Arthur introduces the chalice into the service of the Mass in his kingdom, where it had not been known before. It is as a result of this action, the bringing of the divine blood into the country which has been ravaged by war, that conditions begin to improve.

In most accounts of Perceval's quest a theme of revenge plays an important part, and sometimes ousts the crucial question as the means of healing the Fisher King. The Manessier Continuation of the *Conte du Graal*, for instance, says that the Fisher King had a brother called Gron or Boon (presumably Bron), the King of the Waste Land. He was treacherously murdered by the Lord of the Red Tower, who killed him with a sword which snapped in two. The body and the broken sword were brought to the Grail castle. The Fisher King picked up the pieces of the sword and was instantly wounded in the legs. He could not be healed until a hero came who could rejoin the sword and avenge Gron's death.

Perceval took the sword to a smith named Triburet (Trebuchet in the *Conte du Graal*), who reforged it. The hero used it to kill the wicked Lord of the Red Tower and rode back to the Grail castle with his enemy's head dangling from his saddle. The Fisher King was healed at his approach with the head, which was stuck on top of the highest tower in the castle. It was then revealed that Perceval was the Fisher King's sister's son, and his heir.

Perceval returned to Arthur's court until the Fisher King died. He was then crowned king in the Grail castle, with Arthur and all the knights of the Round Table in attendance, and the Grail served the whole company with food, as with manna from heaven. Perceval reigned for seven years and then retired to the wilderness, taking the Grail and the Holy Lance with him. After ten years he died and the Grail and the lance were never seen again. It was believed that they had been taken up to heaven.

That the Grail is removed from human ken, first to the wilderness and then to heaven, meets the demands of the tradition that Perceval would be the last of the Keepers. If there was to be no one to succeed him, then either he must go on living until the world came to an end (which may have been the original idea) or the Grail must depart from the world. The notion of some great consummation to occur with the coming of the third Keeper has been dropped: perhaps because it was uncomfortably close to the expectation of the Second Coming of Christ. There was a tendency for the Grail quest to become a self-centred mission, whose purpose was more the salvation of the hero than the bringing of any benefit to his fellow men.

The broken sword and the hero's revenge recall Gawain's experiences in the First Continuation. When he came to the Grail castle, at the end of the long causeway leading out to sea, he saw a corpse lying on a bier and was asked to rejoin the two halves of a broken sword. He failed and was told that he would not accomplish the task for which he had come. The task was to avenge the murdered man and so restore the waste land to life (though Gawain did partially restore it by asking about the Holy Lance).

Like the question theme, the revenge story has to do with succession to the kingship. Behind it there seems to lie a tale of the king being murdered and his land falling waste. His heir, the hero, avenges the murder, proves his right to the throne and gives fertility back to the barren country. This merged with the theme of the crippled king, whose land is waste or spellbound, where the hero asks a question which proves his right to the succession and breaks the spell. As a result, a broken sword, or a sword destined to be broken, became part of the furniture of the Grail castle and the hero was frequently committed to a mission of vengeance.

The blending of the two themes sometimes had curious results. In the Welsh story of 'Peredur', in *The Mabinogion*, the hero comes to the fortress of the Lame King, his maternal uncle, and sees a huge spear from which three streams of blood fall to the floor. It is carried across the hall by two young men. Next come two girls bearing a large platter on which is a man's head, drenched in blood. This is an extraordinary metamorphosis of the Grail, but later on it is explained that the head belonged to the hero's cousin, who had been murdered by the nine witches of Gloucester. The nine hags had also lamed the hero's uncle. It had been prophesied that Peredur (Perceval) would take vengeance on the witches and in the end, with the help of Arthur and his men, he kills them. In this story the theme of avenging a kinsman has been mingled with the story of the otherworld vessel and the crippled king, with the grotesque effect that the murdered man's head is paraded through the hall on a plate.

Though the tales of Perceval's quest were constructed largely with pagan building-blocks, their spirit is patently and sometimes militantly Christian. All the adventures of the Grail, according to *Perlesvaus*, occurred to promote Christianity's conquest of paganism, and in the course of the story pagans are triumphantly slaughtered in heaps. *Perlesvaus* is full of pious hermits, who provide a moralistic running commentary on the action by interpreting everything that happens as a lesson in Christian doctrine. The results are sometimes strained, to put it mildly, as in the case of the Questing Beast.

This peculiar creature, as we saw earlier, was usually a symbol of incest, abnormality and evil. It now turns oddly into an image of Christ. Perceval saw it when he rode into the Lonely Forest. It was a gentle, snow-white animal with hounds in its belly which barked and yelped and gave it no peace. Nearby was a standing cross and the Beast ran to it, crouched beside it and gave birth to twelve full-grown dogs. No sooner were they born than they turned on the Beast and tore it to pieces, but they were unable to devour its flesh. The meaning of this scene was afterwards explained to Perceval by a hermit. The Questing Beast stood for Jesus and the twelve dogs for the twelve tribes of Israel, the Jews who were the children of God. They turned on the Saviour and brutally crucified him, but they could not consume the divine body in the Mass. The moral which the hermit drew was that the Jews were and would remain incorrigibly savage and wicked.

Anti-Semitism is endemic in the Grail stories, and so is a pronounced or even fanatical emphasis on the virtue of chastity. The Grail in *Parzival* is attended by virgins of both sexes. The solitary exception is the Grail King, who must marry and have children to continue his line, and Perceval himself is happily married and the father of a son, Lohengrin. In other stories, however, the hero is not merely continent but totally sexless, and *Perlesvaus* says that the thought of loving a woman physically never entered his mind. One reason for this stress on chastity was the hero's resemblance to Christ, and in the background is the body-hating and sex-fearing strain in medieval Christianity. Also, the tradition that the Grail hero would be the last of the Keepers implied that he would have no heir, which in turn could imply that he would have no use for sex.

Enthusiasm for chastity is not incompatible with sexual imagery. On the contrary, the use of erotic symbolism by high-minded writers to convey spiritual truths has a long history. It would not be surprising or shocking if the Grail legends contained imagery of this kind, but how much of it there is in them is not easy to decide. The Grail is the mysterious source of life, physical and spiritual, which means that at one level it is likely to have sexual connotations. The identification of the Grail as a cup may be an example of an erotic symbol used for a spiritual purpose. If so, the temptation is to view the lance in the same light, as the masculine counterpart of the Grail. The scene when the two are carried together in procession with the lance bleeding into the Grail can then be identified, on one level, as a piece of erotic imagery. On the other hand, though a cup is quite a common female symbol, a lance is not so common or obvious a male one and there is no particular reason for thinking it to be one here.

A British scholar, Jessie L. Weston, thought that the cup and the lance had sexual implications. In her book *From Ritual to Romance* (which influenced T. S. Eliot's poem *The Waste Land*), she argued that the Grail legends preserved the rites and symbols of a secret cult, descended from the classical mystery religions and the gnostic sects of the eastern Mediterranean area in the early centuries after Christ. Lying behind the stories of the Grail quest was the ritual of initiation into the cult, which was an initiation into the secret of life. The holy objects of the legends were the sacred symbols of the cult: the dish from which the worshippers ate the communal meal, the lance and cup together standing for male and female as the sources of physical life, and the Grail itself separately as the source of spiritual life. (See also Appendix II.)

This theory fell on academically stony ground, because there is no evidence that the secret cult ever existed, but other writers have suggested that the legends conceal the heretical teachings of the Cathars or similar sects, veiled in hints and cryptic allusions to shield them from the baleful eye of the Church. Miss Weston thought that the Knights Templar may have learned secret gnostic doctrines in the East, hence their connection with the Grail in *Parzival*. Once the Templars are involved, a whole seething cauldron of possible heresy, gnosticism and magic is opened to speculation.

It is clear that the Grail legends could have been affected by heresy and magic. Heresy was so rife in the period when they appeared that in the early thirteenth century the Church organized a ruthless war against the Cathars in the south of France. Unorthodox ideas were in the air, with a common ancestry in the high magic of the later Roman world. Alchemy is one example and there was also an upsurge of interest in astrology. A few *grimoires*, or textbooks of ritual magic, may have been circulating in small numbers. The goal of high magic is to transcend all human limitations and become divine, and this is virtually what happens to the hero who achieves the Grail.

Ideas of this kind do seem to have influenced the Grail stories here and there, in a sporadic way. The concept of the Grail as a stone in *Parzival* is a case in point. But if the legends conceal any organized system of heresy or magic, they conceal it most effectively. The attitude of the Church seems decisive. The Grail stories cold-shouldered the Church, because they were fundamentally heroic tales in which the central character drove his way to success by his own efforts. The Church treated them with reserve. If there had been grounds for suspecting that they were a cover for heresy or magic, the Church would have been after them like a ferret down a rabbit-hole. These were not isolated alchemists or magicians, working in secret, but part of the most popular literature in Europe. The fact that no campaign was mounted against the legends strongly suggests that there was no systematic pattern of deviationist ideas in them to be detected.

The most damaging attack on the legends came from within, as it were, when the Grail quest was used as a vehicle for Cistercian propaganda. The Cistercians, or white monks, reached the height of their influence in the early thirteenth century. Their ideals of purity, austerity, simplicity and the disciplined dedication of the soul to God were injected into a new version of the quest in the Vulgate Cycle, which imposed order and coherence on the whole rambling Matter of Britain. The winning of the Grail now became the supreme moment in the history of Arthur's kingdom and a brand-new Grail hero was invented, who was free of the worldly and semi-pagan atmosphere clinging to Perceval. The new hero was Galahad.

The Triumph of Galahad

The way was prepared for Galahad in the Vulgate *Lancelot* and the story of his triumph was told in the *Queste del Saint Graal*. The Grail is here identified as a dish. It is the platter from which Jesus ate the Passover lamb with his disciples, so that it is related to both the Last Supper and the symbolism of Christ as the sacrificial Lamb of God. As before, however, the Grail is associated with luxurious feasting and as before it changes shape. It is seen as a chalice and as a ciborium. At one point the blood running down the shaft of the Holy Lance is caught in it.

There is still no question of bringing the sacred relic within the control of the Church. It is guarded by its hereditary Keepers, who draw their authority from a higher source than Rome. It is carried by a beautiful woman, the Grail King's daughter, not by a priest. The heroes in quest of it are now helped by Cistercian monks as well as hermits, but Galahad does not win the Grail by way of the Church. He achieves it through the direct favour of God.

The Grail, we are now told, was brought to Logres, or Britain, by Joseph of Arimathea himself. He came with a small army of followers, four thousand of them or more, and many of them lived without food through the power of the Grail. It was kept in the castle of Corbenic by a succession of kings, descended from Joseph. The name Corbenic may be a corruption of *cor benoit*, 'blessed body', meaning the body of Christ, which Joseph took down from the cross, and referring to the Divine Presence in the stronghold of the Grail. The castle retains its otherworldly habit of appearing and disappearing. Knights who have been there cannot rely on finding it again and even Galahad, born and bred at Corbenic, takes a long time to reach it in the quest.

The great-great-grandfather of Galahad was King Lambar or Labran (which seems to preserve the name of Bran). He was killed in battle, split in two by a terrible blow from a sword which had mysteriously arrived in the country and was later discovered to be the sword of King David. The blow which killed Lambar turned Logres into a waste land. No crops would grow, no trees bore fruit and the fish vanished from the streams. The enemy who struck the blow fell dead, for no man could draw the sword of David without being killed or maimed, until the coming of Galahad, for whom it was destined.

Lambar was succeeded by his son Parlan, who happened across the fateful sword, drew it partly from its sheath and was instantly pierced through the thighs by a flying lance, propelled by no visible hand. Parlan never recovered from the wound. Known as the Maimed King, he dragged out a pain-wracked existence behind the scenes at Corbenic. His son Pelles, the Fisher King, was the master of Corbenic and custodian of the Grail in Arthur's time, and Pelles in turn was the grandfather of Galahad.

The Fisher King and the Maimed King have now become two different characters and the waste land is connected with yet a third. Nothing more is said about the waste land, but part of Galahad's mission is to end the enchantments of Britain, to break a spell that is

Family Tree of Lancelot and Galahad

According to the Vulgate Cycle and Malory

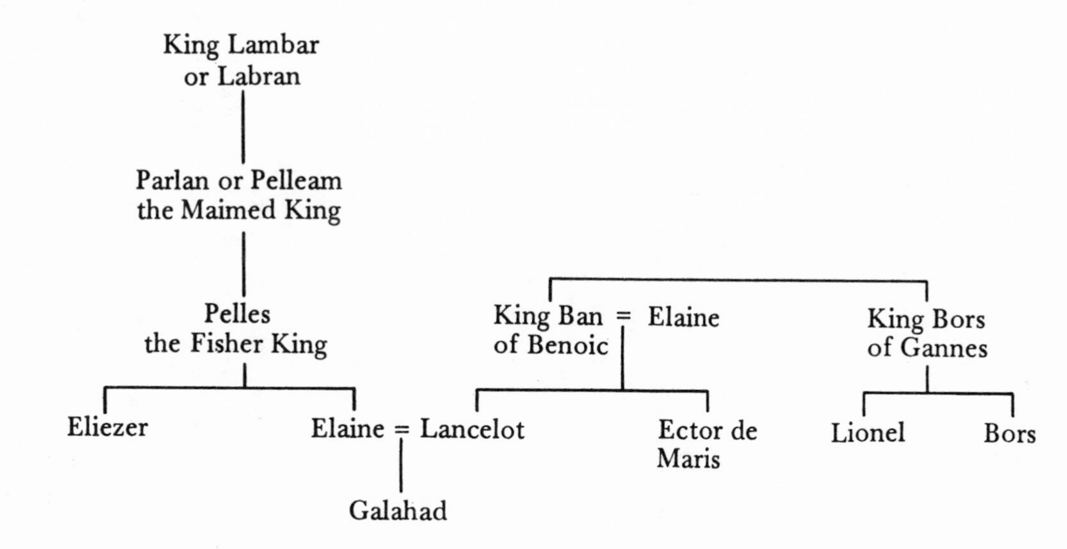

lying on the land. He is also to heal the Maimed King, but not by asking a question. The theme of the crucial question is dropped altogether, and with it the episode of the hero going to the Fisher King's castle and failing to ask about the Grail. Galahad is above failure.

Galahad is the son of Lancelot and the Grail Bearer, Elaine, the beautiful daughter of King Pelles. He is conceived when Lancelot comes to Corbenic and, with the connivance of Pelles, is cast under a spell in which he makes love to Elaine in mistake for Guinevere. This episode has been fiercely attacked by critics for tarnishing the purity of the Grail Bearer and making the good Fisher King a pander for his own daughter, but there is a brilliant irony in Lancelot fathering the pure Grail hero while in his mind making adulterous love to Guinevere, which is the sin that will debar him from winning the Grail himself.

The importance of giving a great hero a conception outside marriage and in circumstances of mystery and magic evidently outweighed the demands of conventional propriety. As the son of Lancelot, Galahad is the offspring of the best knight in the world, and he will be even greater than his father. Lancelot stands for Adam, the imperfect man, and Galahad for Christ as the Second Adam, or man perfected. The sin which Adam committed in Eden, the sin which brought death into the world, is identified as lust. Desire is the root of death, virginity is the passport to immortality, and Galahad will be virgin all his life.

The story confers on Galahad the hereditary right to the Grail which had previously belonged to Perceval. It gives him a double descent from Joseph of Arimathea, through the female line on both sides, because Elaine and Lancelot's mother were both descended from Joseph. Through Lancelot, Galahad's lineage goes back to King David, who was an ancestor of Christ. His name in French is Galaad, from Mount Galaad or Gilead in the Bible, and the interpretation of the name in Genesis as 'the heap of witness' was believed to be a reference to Christ.[21]

The young Galahad was brought up without a father and was presently sent away to a nunnery, where he was trained by women. In these ways he resembles both Lancelot and Perceval. He remained at the nunnery until the time came for him to take his place at the Round Table, the most exalted and most dangerous place, in the Siege Perilous.

One year at Pentecost a marvel was seen at Camelot. A great stone, of red marble, came floating down the river and drew in to the water's edge. In it was fixed a superb sword, with an inscription on the pommel which said that it could only be taken from the stone by the best knight in the world. Arthur thought this must mean Lancelot, but Lancelot said he was not worthy of it and he would not attempt to take it. Gawain modestly declined to try, until at Arthur's command he tugged at the sword, but in vain. Then Perceval tried, but he could not move it.

Arthur and the knights returned to the royal palace and sat down at the Round Table to eat. Suddenly all the doors and windows in the palace closed by themselves. An old man in white robes appeared, and with him a young knight who wore red armour but carried no sword. The old man announced that this was the champion who would break the spell on the land. The young knight was Galahad and the old man led him to the Siege Perilous. Galahad sat down in it. Everyone held their breath, but nothing untoward happened, and the news ran round the court that the knight who was to achieve the great adventure had come. After the meal, Arthur took Galahad to the waterside, where he drew the sword from the marble stone and girded it on.

In the evening the knights gathered again at the Round Table. There was a fearsome clap of thunder, a bright ray of sunlight shone into the hall and all the knights were struck dumb and gazed at each other in wonderment. Then there came into the hall through the great door the Holy Grail, borne by no mortal hand, and the air was filled with fragrance and every man had the food which in all the world he most desired. What the form of the Grail was, however, no one could see. The Grail vanished and they all recovered the power of speech.

The episode of the sword in the stone echoes the story of how Arthur had established his right to his throne years before. It demonstrates, before Galahad has even arrived at Camelot, that he is the best knight in the world, the superior of all others, including not only Lancelot and Gawain but the former Grail hero, Perceval. This is driven home by Galahad's successful occupation of the Siege Perilous, which here represents the seat of Jesus at the Last Supper, not the seat of Judas, so that a parallel between Galahad and Christ is again implied.

The Grail retains its old connection with delicious food and its old ability to tie the beholder's tongue. Its appearance at the Round Table at Pentecost is related to the experience of Christ's disciples at the first Pentecost, when the Holy Spirit descended on them in tongues of fire. Galahad's red armour and the red marble stone are symbolically related to the fire of the Holy Spirit. Far from being struck dumb, the early Christians received the gift of speaking in tongues. Arthur's knights, by contrast, are sinners. They have been shown a wonder, but they have received no spiritual gifts.

After the Grail had left the hall, Gawain spoke up. He had earlier been to Corbenic and seen the Grail, but his attention had been distracted by the beauty of the Grail Bearer and he had been driven ignominiously from the town. He now swore that he would go in search of the Grail and not return to court until, if he was worthy to do so, he had seen it plain. Lancelot, Perceval and all the other knights also vowed to learn the truth about the Grail. Arthur was deeply grieved, for he knew that he would never see his whole company of the Round Table at his side again. Gawain now wished he had kept silent, but he could not withdraw his pledged word. Guinevere wept bitterly because she would be parted from Lancelot, and so did the other ladies when they discovered that no knight could take a woman with him on the quest of the Grail.

The knights rode out with high hopes from a court dissolved in tears but, apart from Galahad himself, only three of them achieved any measure of success: Lancelot, Perceval and Bors, who was Lancelot's cousin. Gawain and the rest were too worldly and too sinful to realize that they had undertaken a spiritual journey, not a search for exploits of the kind to which they were accustomed. They complained that they found no adventures at all, still less any sign of the Grail, and sooner or later they drifted back to Camelot.

Lancelot, Perceval and Bors had various supernatural encounters, saw strange visions and were plagued by elaborate demonic temptations, the Christian symbolism and significance of these experiences being expounded to them with excessive zeal and loquacity by a swarm of hermits. These pests clutch the reader's sleeve from every other page. The story almost suffocates in a fog of sententious piety.

The emphasis on chastity now becomes an obsession. Lust is the root of all sin and raises the drawbridge of heaven. Perceval's qualification for success in the quest is that he is a stranger to sexual desire.

He and Galahad are the only virgin knights of the Round Table; Bors has slept with a woman once but has sincerely repented of it. Perceval's besetting fault, however, he shares with Lancelot and Arthur's other champions. In his arrogant confidence in his rank and his strong right arm, he relies on himself instead of God.

As far back as the *Conte du Graal*, excessive self-reliance had been criticized, but the criticism is now much more fundamental. The earlier principle of trusting in God and keeping one's powder dry will no longer do. The message of Galahad's triumph in the *Queste del Saint Graal* is that no one, however admirable, can win his own way to God. The gap between God and sinful humanity is far too wide for that. Only the grace of God, freely given and undeserved, can bring man to God. The *Queste* is not a hero-tale but a polemic, and its target is the ideal of knightly chivalry, which essentially valued a man to the extent that he depended not on God, but on himself.

This is why the motif of the crucial question has been dropped. The question implies the principle of 'Ask, and it will be given you; seek, and you will find.' This principle is negated in the *Queste*. Some of the earlier stories had already suggested that the hero could not make his way to the Grail castle, and so to the ultimate spiritual achievement, by earnestly seeking it. He must come there by chance or through the will of God, which amounted to the same thing. Not much emphasis had been placed on this idea, however. Now the favour of God is all-important and without it the seeking is vain. By opposing a godly ideal to a manly one the *Queste* attacked the whole spirit of the older Arthurian romances. When one of the hermits lists the virtues in order of merit, the first three are virginity, humility and patience. They are scarcely the virtues of the Arthurian heroes of yore.

As the finest representative of the older spirit, Lancelot is given a stony path to tread. The nobility of his character is recognized, but he cannot succeed in the quest because of his adultery with Guinevere. When this is impressed on him, he confesses his sin and swears to give her up. He listens in silent dejection while she is reviled as a worthless creature who does not really love him and who has robbed him of his hope of heaven. He does humble penance and dons a hair shirt. Humiliated and robbed of his self-confidence, mocked because his own son will outdo him, he hews miserably to the quest. He gains merit by taking counsel of the hermits at every stage, instead of fending for himself, and eventually he is allowed to see the glory of the Grail from a distance. When he tries to go closer, a flame seems to shoot through him and he is knocked unconscious.

Galahad is a cardboard character and the dice are so heavily loaded in his favour that it is difficult to feel any enthusiasm for his success. Because he has no failings, he is not human. The parallel with Christ is constantly though unconvincingly drawn and one of the ubiquitous hermits tells Galahad that his coming must be compared to the coming of Christ, adding hastily 'in semblance only, not in sublimity'. Since he is perfect to begin with, he cannot change or grow in any way, and so there is comparatively little he can be given to do. He liberates the prisoners in the Castle of Maidens, which is compared to Christ's freeing souls from hell. He defeats all comers with matchless prowess at a tournament, which represents the victory of good over evil. He finds and draws the fell sword of David.

In the end Galahad, Perceval and Bors joined forces and came to Corbenic, where they were joyfully welcomed by King Pelles. A mysterious voice summoned them to sit at the table of Jesus Christ and feed on the food of heaven. The old Maimed King was brought into the room on a litter. Then the three knights saw the Holy Lance and the Grail, carried in procession by angels, with the blood flowing from the lance into the Grail. Kneeling before a silver table on which the Grail was placed, they saw Christ appear from the sacred vessel, naked and bleeding from the hands and feet. Christ gave them communion from the Grail and they found this holy food 'so honeyed and delectable that it seemed as though the essence of all sweetness was housed within their bodies'.[22]

Christ told them that the Grail was to leave Logres, because the people of the country did not honour it as was its due and had become sinful and worldly. With its departure the spell lying on the land would be broken. When Christ vanished from their sight, Galahad took blood from the Holy Lance on his fingers and touched the legs of the Maimed King. The king was instantly restored to vigorous health and gave thanks to God. He spent the rest of his days in a Cistercian monastery, where God performed many miracles for love of him.

That night the three companions left Corbenic and rode to the sea, where they found the Grail and the lance awaiting them in a ship which sailed by itself. The ship took them to Sarras (Jerusalem), where the citizens made Galahad king. A year to the day after his coronation, Galahad went with Perceval and Bors to the place where the Grail was kept. They saw a man dressed as a bishop celebrating the Mass of the Virgin Mary. Presently he called Galahad to look into the Grail. When Galahad looked into it, the innermost mystery was revealed to him. Trembling violently, he raised his hands to heaven and said: 'Lord, I worship Thee and give thanks that Thou hast granted my desire, for now I see revealed what tongue could not relate nor heart conceive. Here is the source of valour undismayed, the spring-head of endeavour; here I see the wonder that passes every other.'[23]

After this revelation Galahad no longer had any desire for life on earth and angels bore his soul to heaven. Perceval and Bors, who had not shared the ultimate experience, saw a hand come down from the sky, take the Grail and the Holy Lance and carry them up to heaven. Sorrowfully, they buried Galahad in Sarras. Perceval became a hermit and died a year later. Bors made his way back to Camelot, where he was greeted with great rejoicing. The Grail and the lance were never seen again.

So the Grail has departed, and with it the enchantments and strange adventures which the presence of this holy thing in the human world had induced. That it is taken away by Christ, to punish the people, strikes a characteristically sour note. The secret of the Grail is something that 'tongue could not relate nor heart conceive'. This is a reference to St Paul: 'What no eye has seen, nor ear heard, nor the heart of man conceived, what God has prepared for those who love him.'[24] The Grail has reached its apogee in the myth as the channel of union with God, through which is attained on earth the Beatific Vision, the experience of seeing God face to face. With this experience comes the realization that in God alone, and not in man, is the source of valour and the spring-head of endeavour.

The purpose of the quest is no longer succession to a throne, unless it is a heavenly one. Healing the crippled king and breaking the spell on the land are now only incidental to it. Nor is it the achievement of integrity and the true self. Galahad does not find his ideal self in the quest, because he is perfect to begin with. He is consequently the only knight who receives the grace of God in full measure. Even the best of the others, hard though they try, do not succeed completely. They do not achieve the ultimate experience and they do not learn the whole truth about the Grail, as they had set out to do. They have to be content with second-best: a splendid second-best to be sure, but still comparatively a failure.

As before, to achieve the Grail is to achieve immortality, but the quest is now an almost totally self-centred expedition. Galahad is not a hero in the tradition of those who confer some inestimable blessing on their fellow men. In the rejoicing which greets Bors when he returns to Arthur's court, there is perhaps a faint echo of the 'great joy' which Gawain was told would attend success in the quest, but it is extremely faint. The quester's primary duty is to himself alone, to the welfare of his own soul. When he succeeds, the Grail is taken away from the world and, although this lifts the spell from Logres, it is represented as a punishment. For all the parallels industriously drawn between the hero and Christ, Galahad's triumph brings benefits to few except himself.

It remained for the author of the *Suite du Merlin* to put together, at last, a satisfactory version of the Waste Land theme. A knight named Balin, who was doomed to misfortune, went to the land of Listinois, which was ruled by King Pelleam, the worthiest and most Christian king alive. Pelleam had a sinister brother, whose face was black and who rode about invisibly, murdering knights with a spear. Balin killed this evil and shadowy figure, and was at once attacked by Pelleam. In the fight Balin's sword was broken. Pursued by the vengeful Pelleam, he ran from room to room of the castle, looking for a weapon, until he came to a chamber which a mysterious voice told him not to enter because it was a holy place. Taking no notice, Balin rushed in. It was a room unrivalled in the world for beauty. In it was a massive silver table, on which stood a vessel of silver and gold, and in the vessel, hanging unsupported and head downwards, stood a lance. A voice told Balin not to touch it but, desperate and with Pelleam hot on his heels, he seized the lance and ran the king through the thighs. The walls of the castle fell in, Pelleam collapsed and Balin dropped as if pole-axed. A loud voice proclaimed that 'the High Masters' would take vengeance on those who had not deserved it for Balin's sin in touching the Holy Lance and wounding Pelleam. The marvels and adventures of Logres would now begin.

When Balin came round, he found the people of the castle hurt by falling debris, struck dead by the terror of the voice or in a state of helpless panic. Riding away from the castle, he found the country blighted, crops destroyed, trees withered and people lying dead in the fields and the towns. From that time on, Listinois was called the Kingdom of the Waste Land. The doomed Balin was soon afterwards killed by his own brother. King Pelleam was severely wounded and had to wait many years for the coming of Galahad to heal him.

Ingredients from earlier stories are here combined in an effective pattern. The lance standing in a vessel of silver and gold is the lance Gawain saw bleeding into a silver vessel at the Grail castle in the First Continuation. The invisible enemy who strikes with a spear is related to the Grail King. The motif of the broken sword is fitted in. The blow which maims the Grail King is also the Dolorous Stroke, the devastating blow which creates the waste land. Previously struck by a spear or a sword or a mysterious flying lance, it is now dealt by the Holy Lance itself. The Grail King is wounded in the thighs, which implies emasculation. His land is instantly blighted and the enchantments of Britain begin at this moment, though this is not an automatic magic consequence of the maiming of the king but a punishment inflicted by higher powers for Balin's sacrilege. Eventually, Galahad will heal the Maimed King and the enchantments will end. The unhappy Balin, whose steps are dogged by tragedy and who unintentionally causes the disaster, is the opposite and counterpart of Galahad, the favourite of God, who will repair it.

It seems odd that the good King Pelleam's castle should harbour his sinister brother, who is a treacherous murderer and in whom it is possible to see a dark twin or aspect of Pelleam himself. It is strange that the Dolorous Stroke which blights the country and casts a spell on Logres should be struck by the Holy Lance, or similarly in the *Queste* that the sword of David should wreak such havoc. The legends, however, preserve the old sense of the intensely harmful potential of holy objects, people and places. The Grail itself, which is dangerous to talk about, has the same quality. A thing is not sacred because it is good. It is sacred because it contains mysterious and awesome power. It is as potent for good or evil as a huge charge of electricity. If it is tampered with, however compelling and understandable the motive, the consequences may be catastrophic for entirely innocent people.

In the Christian myth of the Grail the hero achieves his ideal integrity of character as the best knight in the world: indeed in the *Queste* he achieves *the* ideal integrity in becoming one with God.

The hero discovers the secret of life and by winning immortality he triumphs in the ultimate battle against death. He ought to return to the human sphere with the gift of immortal life for his fellow men, but this cannot be allowed him because it would anticipate the Second Coming of Christ, the Last Judgement and the end of the world. The difficulty with the myth was the danger of the hero usurping the position of Christ. Medieval writers found no satisfactory solution and in the end, when the myth reaches its clearest formulation in the Vulgate Cycle, the hero has to be stripped of all humanity and turned into an unconvincing effigy of the Saviour. At the same time the Grail has to be removed from the world to deny any possibility of a human hero achieving it. The legends which grew up about Glastonbury constitute, in effect, a protest against this pessimistic conclusion.

The Glastonbury Legends

Though the Grail is taken away from the human world altogether in the *Queste*, and in several of the other stories, a popular legend grew up that it was hidden at Glastonbury. The refusal of the authorities at Glastonbury Abbey to adopt any such belief or to link their community directly with the mysterious vessel is a striking example of the coldness with which the Grail was treated by the Church.

The Grail is an element of Glastonbury's strange, magnetic, spellbound atmosphere. The ruins of the abbey, themselves powerfully evocative, lie near the foot of Glastonbury Tor, a conical hill which rises steeply from low-lying ground and can be seen for miles around. Until the surrounding marshes were drained in the middle ages, the Tor rose from what was virtually an island in a wilderness of tangled swamps. Long before the Romans came to Britain, a thriving village stood on an artifical platform in the marsh, a mile to the north of the present town. It was a prosperous community, with an unusually high standard of culture and trade links with Wales, Ireland and Brittany.

The village was destroyed by raiders in the first century AD, but memories of a sanctuary of civilization in the wilderness may have lingered on to influence the tradition of Glastonbury as a special place, set apart from the ordinary everyday world. The Celts connected islands, hills and mounds with the otherworld, and Glastonbury Tor was an impressive hill on an island. The identification of Glastonbury as the otherworld Isle of Glass or Isle of Avalon probably goes a long way further back than the twelfth century, when we first hear of it. There was a folk belief, recorded in the twelfth-century biography of St Collen (a Welsh hermit of the seventh century) that Gwynn ap Nudd, the lord of Annwn, had a palace on top of the Tor. According to the story, St Collen, who had built himself a small cell at the foot of the hill, was summoned by a mysterious voice to meet Gwynn. Taking plenty of holy water with him, he climbed the Tor and saw on the top a beautiful castle, where Gwynn sat enthroned in a golden chair, surrounded by retainers in red and blue livery, minstrels making music, handsome young men and lovely girls. On a table were the most luxurious delicacies imaginable and Gwynn hospitably offered the saint something to eat. Collen declined, sensibly, because to eat the food of the otherworld is to risk becoming a prisoner there. He sprinkled his holy water in all directions and the castle and all its people vanished. He found himself alone on the top of the hill. It may be significant that when a chapel was built on top of the Tor it was dedicated to St Michael, the vanquisher of Satan and the demons.

Another old element of Glastonbury's spell is the tradition that it was the oldest Christian foundation in Britain, the place where the earliest Christian missionaries settled. In the grounds of the abbey stood a small building of wattle and daub, which was believed to be the first church ever built in the British Isles. This was the Old Church, dedicated to the Virgin Mary, which was burned down in the disastrous fire of 1184. It was said that the disciples of Christ erected the Old Church with their own hands, which would date it to the first century AD.

William of Malmesbury, who researched Glastonbury's history in the 1130s, doubted if the Old Church was quite as old as that. He thought it had been built about 180, when missionaries were sent to Britain by Pope Eleutherius. However, he conceded that Christianity might perhaps have come to Glastonbury earlier still. There was a tradition that the Apostle Philip preached in France soon after the Crucifixion and, if this was true, it was possible that he had planted the seed of the gospel on the British side of the Channel as well.

The discovery of the graves of Arthur and Guinevere at the abbey in 1191 appeared to clinch the identification of Glastonbury with Avalon and to link the place firmly with Arthur. Writing soon after the discovery, Robert de Boron had the Grail taken to the West. He also sent many of his characters, though not Joseph of Arimathea himself, to preach Christ in the West and one of them, named Peter, was to go to 'the vales of Avaron' and await the coming of the third Grail Keeper. Robert's characters appear to be the first Christian missionaries to Britain, and Avaron or Avalon almost certainly means Glastonbury. The author of *Perlesvaus*, a few years later, claimed to have taken his story of the Grail from a book in the library of Glastonbury Abbey. The claim may not have been true, but there was evidently a tendency to hint at a close connection between Glastonbury and the Grail.

None of the romances, however, puts the sacred relic itself at Glastonbury. The First Continuation says that Joseph of Arimathea came to Britain himself, bringing the Grail with him, and so do the *Queste* and the *Estoire del Saint Graal*. The *Estoire* gives Joseph a leading role in the conversion of Britain to Christianity and sends Peter to the Orkneys, where he converts the king, marries the king's daughter and becomes the ancestor of Gawain. But in all this there is no mention of Glastonbury.

By about 1240 the Glastonbury monks themselves had produced the following story. The Apostle Philip in France sent twelve of his disciples to Britain, led by his friend Joseph of Arimathea. They arrived in AD 63. The local king turned a deaf ear to their attempts to convert him, but allowed them to settle on the Isle of Glass, isolated among the swamps. There they lived as hermits, spending their time in fasting and prayer, and there they built the Old Church in honour of the Virgin Mary. When the last of them died, the place remained deserted until the missionaries sent by Pope Eleutherius came to Glastonbury, restored the Old Church and founded a new community of twelve hermits, from which the medieval monastic community was descended.

Leaving Joseph of Arimathea aside, there is nothing far-fetched in the idea of a small group of recluses at Glastonbury at an early date. The Celts did not lose their reverence for islands when they became Christians. On the contrary, Celtic Christian hermits established themselves on islands and in other remote places where, far from the madding crowd, they could feel close to God. Glastonbury would have been an ideal site. It is now thought that the monastic community at Glastonbury was founded by Irish monks early in the sixth century, perhaps in the real Arthur's time, but that a shrine of some kind may have existed there already.

The official legend says nothing about the Grail. The abbey laid no claim to that fascinating but unorthodox talisman, although it did claim to possess numerous relics of the Passion, including several pieces of the cross, some of the earth in which the cross stood, a thorn from the crown of thorns, and part of the table of the Last Supper. What the legend does say is that Joseph of Arimathea was the founder of Christian Glastonbury. This is interesting, because Joseph owed his fame to the Grail romances as the first Keeper of the Grail. It looks as if the authorities at Glastonbury wished to attract this valuable aura of mysterious prestige to the abbey, but without actually claiming the Grail.

Later on, as a substitute for the Grail, the monks said that Joseph had brought with him two small silver vessels, one containing the blood and the other the sweat of Christ on the cross. These vessels were buried with Joseph in his grave at Glastonbury, but nobody knew where his grave was.

Once Joseph of Arimathea had been authoritatively linked with Glastonbury, popular appetite created a legend that placed the Grail there as well. How long ago this legend evolved is not clear, but we have seen authors hinting at a connection between the abbey and the Grail early in the thirteenth century. In its complete form the story is that Joseph and a small band of missionaries came to Britain thirty years or so after the Crucifixion. With them they brought the Grail, the cup of the Last Supper. Landing somewhere in the south-west, they struck across country until they came close to the foot of Glastonbury Tor. There they stopped to rest and pray, and Joseph thrust his staff into the ground. It immediately took root and put out buds. It was the ancestor of the famous Glastonbury Thorn, which flowers every year at Christmas: or, since the reform of the calendar in the eighteenth century, in January. Joseph and his companions took the miracle as a sign that they had reached the end of their journey. They settled at Glastonbury and built the Old Church. To preserve the Grail from profane hands, Joseph buried it somewhere at the foot of the Tor. The spring now known as Chalice Well is sometimes identified as the place.

According to a variant of the story, when Joseph and the missionaries came to Glastonbury, they found the Old Church already standing, 'built by no human hands'. It had been constructed by Jesus himself, who was brought up to the carpenter's trade and had visited Somerset years before. Hence the opening lines of Blake's 'Jerusalem':

> And did those feet in ancient time
> Walk upon England's mountains green?
> And was the holy Lamb of God
> On England's pleasant pastures seen?

A recent elaboration of the legend makes Jesus the nephew of Joseph of Arimathea, sent to Britain by his rich uncle to gain experience of the tin trade. This is supposed to explain why the New Testament contains no information about Jesus's late boyhood and early manhood.

The legend about Jesus in England is pleasing, improbable and unsupported by evidence. So is the legend of Joseph of Arimathea in Glastonbury. The celebrated Glastonbury Thorn, a variety of hawthorn which blossoms in winter, is apparently a comparatively late addition to the story. The earliest reference to it dates from about 1500. The tale about Chalice Well seems to have sprung up in the nineteenth century. The possible kernel of truth in the legend is that Glastonbury may have been a Christian site from a very early date (and perhaps a pagan religious site of some kind before that).

Another interesting variant is that the Grail was not buried, but was kept safe in secret by the Glastonbury authorities. When Henry VIII forcibly closed the abbey in 1539 and the last abbot was executed on the Tor, the Grail was smuggled away to the small Cistercian abbey of Strata Florida in Cardiganshire, far off in deepest Wales. This abbey was closed in the same year and its few monks were sheltered by the Powell family of Nanteos near Aberystwyth. The last survivor of the monks gave the Grail to the head of the family for safe keeping, telling him that a drink from it would cure disease. It was kept at Nanteos for generations. It was not a golden chalice studded with precious stones but a small, simple, wooden vessel, worn with age, such as the cup of the Last Supper might actually have been. Its healing powers were well known in the district and sufferers from numerous diseases were cured by water which had been poured from it.

There is no doubt that a healing cup was long kept at Nanteos, but there is an obvious difficulty about the story that it was the Grail and came from Glastonbury. If so, why did the Glastonbury authorities pretend to know nothing of it? They made no secret of their other sacred relics – quite the contrary. A tentative answer might be that they regarded the Grail as something so uniquely holy that its existence must not be revealed, and that they were also influenced by the Church's hostility to the Grail legends. The answer is not very convincing and one is driven reluctantly to the conclusion that the cup of the Last Supper was never brought to Glastonbury and did not survive at all. The Grail is a haunting and magnificent fiction. What the Glastonbury legends really bear witness to is two things. The first is the uncanny and spell-binding atmosphere of the place, which gave Glastonbury so long ago an air of special meaning, of being an enclave of mystery and significance beyond the normal experience of man, the otherworld Isle of Avalon. The second is the feeling that the Grail must not be taken away from the world, which would be too impoverished without it. Somewhere on the edges of the known world, somewhere on the borders of the conscious mind, the secret of life is guarded still.

4
The Passing of Arthur

> Alas, we know very well that Ideals can never be completely embodied in practice. Ideals must ever lie a very great way off; and we will right thankfully content ourselves with any not intolerable approximation thereto! Let no man, as Schiller says, too querulously 'measure by a scale of perfection the meagre products of reality' in this poor world of ours. We will esteem him no wise man; we will esteem him a sickly, discontented, foolish man. And yet, on the other hand, it is never to be forgotten that Ideals do exist; that if they be not approximated to at all, the whole matter goes to wreck!
>
> Carlyle, *On Heroes and Hero-Worship*

When Galahad came to Camelot and took his place in the Siege Perilous, the magic circle of the Round Table was complete. It was soon broken again, for the knights rode out on the quest of the Grail, from which Galahad and Perceval did not return. Arthur's sorrow and foreboding when all his champions vowed to go in search of the Grail was fully justified. Once the Grail had been won, the Round Table was doomed. The supreme objective for which it was founded had been achieved. The wave had crested and now could only break.

This last part of the story, however, is no anti-climax but the most moving section of the whole Matter of Britain. It is magnificently told in the Vulgate *Mort Artu* (on which Malory based his own superb account of the ending of Arthur's kingdom). In these final scenes Arthur, Lancelot and Gawain come back to the centre of the stage. Lancelot's passionate relationship with Guinevere proves to be the rock on which the fellowship of the Round Table is splintered and destroyed. Caught in a tangle of treachery and conflicting loyalties, Arthur goes down fighting to the last.

In the *Mort Artu*, released from the world-rejecting piety and incessant sermonizing of the *Queste del Saint Graal*, the story returns to the real world and the eternal verities of human nature: love and hate, jealousy and generosity, pride and power-hunger, honour and shame. Through an intricate web of circumstances, human emotions bring the Matter of Britain to its tragic close.

Lancelot and Guinevere

During the quest of the Grail, Lancelot had sworn to forsake Guinevere. Within a month of returning to Arthur's court, however, his love for her burned as ardently as ever and they soon became lovers again. They were now so indiscreet that Agravain, Gawain's brother, discovered their intrigue. Agravain had never much cared for Lancelot. Pleased with the chance to injure him, he took his information to Arthur, who refused to credit it. He told Agravain that Lancelot would never have acted in such a way. If he had, he must have been driven by the overwhelming force of love, against which reason and common sense were helpless, but no, Arthur could not believe it of him. Significantly, Arthur did not say that he could not believe it of Guinevere.

Not long after this, Arthur and some of his retainers were travelling through a forest and lost their way. They came to a castle they had never seen before, where they were hospitably welcomed. Equipped with a splendour which Arthur himself could hardly rival, the castle was the stronghold of Morgan le Fay. When Arthur met his half-sister again, for the first time in many years, he was generously overjoyed. With a blind disregard of reality, he asked her to go back to Camelot with him and take up her old position as lady-in-waiting to Guinevere. Morgan politely declined.

Morgan was determined to open her brother's eyes to the truth about Guinevere and Lancelot, both of whom she still fiercely detested, but she feared that nothing would stop Lancelot from killing her if she did. She solved the problem by giving Arthur the room in which she had kept Lancelot prisoner, the room on whose walls he had painted pictures of his own career. As a painter, evidently, Lancelot had no great gift for realism, for he had identified the scenes and characters in captions. The pictures and captions left little doubt that Lancelot and the queen had been lovers. Arthur cross-examined Morgan and with feigned reluctance she told him that he had been deceived. She urged him to punish the guilty pair, because no true king or true man could live with such dishonour.

Arthur went back to Camelot, to find that Lancelot was away from court, no one knew where. The king began to doubt whether Morgan had told him the truth. One of his most engaging characteristics was always to think the best of people. He remained uncertain, suspicious and deeply uneasy.

Lancelot, meanwhile, was at a place called Escalot (Astolat in Malory, who identifies it as Guildford), where he was slowly recovering from a severe wound received in a tournament. He was nursed by a beautiful girl, who fell helplessly in love with him. Tortured by her longing for him, she begged for his love, but he told her that his heart was given elsewhere. The unfortunate girl took to her solitary bed, pined away and died. At Camelot, ironically, Guinevere believed that Lancelot was so long away because he was having an affair with another woman. Furious with jealousy, she told Bors, who was Lancelot's cousin and loyal friend, that she hated Lancelot even more than she had ever loved him.

Gawain and several other knights were eating at the queen's table one day when a poisoned fruit, intended for Gawain by an enemy who had a grudge against him, was given to Guinevere. She innocently offered it to one of the knights, who bit into it and fell dead at the table. Everyone at court was horrified and jumped to the conclusion that Guinevere had murdered the knight. Arthur himself, already suspicious of her, doubted her innocence. The dead knight's brother, Mador de la Porte, went to Arthur and demanded justice. The only way in which the queen could clear herself was to find a champion to defeat Mador in trial by combat, but despite her entreaties no one at Camelot would fight for her, because they all believed her guilty. Arthur gave her forty days in which to find a champion.

At this point a boat, richly hung in silk, came floating down the river to Camelot. Arthur and Gawain went aboard to investigate and found that the boat bore the dead body of the maiden of Escalot. In a purse at her belt was a letter explaining that she had died of unrequited passion for Lancelot. When Guinevere heard of it, she reproached herself bitterly for her suspicions of her lover. She still did not know where Lancelot was and she was in terror of her life. Arthur, who still loved her, was trying hard to find her a champion against Mador de la Porte, but none of the knights, even Gawain, would consent. They refused to fight in a wrongful cause.

Lancelot was still away when he heard what had happened at court. Though he thought that Guinevere must be guilty, he resolved to fight for her on the appointed day in gratitude for the admiration and respect she had shown him throughout his career. Since he would be defending an unjust cause, however, he decided not to fight quite as hard as was his wont.

We are given no convincing reason for the readiness of everyone, including Lancelot, to believe Guinevere capable of a treacherous murder, nor had she any motive for killing the knight. Either this casts a new light on the queen's character or it is an example of the age-old masculine belief in the subtle and incomprehensible evilness of women.

The day of the trial by combat came and Lancelot, who had returned to Camelot secretly, rode on to the field incognito to defend the queen's cause against Mador de la Porte. At the first collision Lancelot unhorsed Mador. He then dismounted, for he considered it unchivalrous to fight an opponent on foot from horseback. The duel continued until Mador was cut and bleeding in a dozen places and Lancelot, not wanting to kill him, asked him to withdraw his accusation against the queen. Mador realized that he was facing Lancelot and gratefully accepted the offer. Arthur and Gawain ran to embrace Lancelot, whose victory showed that Guinevere was not guilty.

Lancelot returned to Guinevere's bed and whispers of their adultery ran round the court. Not only Agravain, but all the Orkney brothers now knew of it. Arthur found them talking about it one day and angrily demanded to know what they were saying. Gawain and Gareth, who greatly admired Lancelot, flatly refused to answer, but Agravain and Mordred told the king that he was being cuckolded. Arthur ordered them to catch the lovers together.

The unfortunate king was at the same time reluctant to believe the accusation and anxious for clear proof of it. He had also to consider the fact that if he had Lancelot executed, Lancelot's redoubtable kinsmen – his brother Ector de Maris and his cousins Bors and Lionel – would bring a war round Arthur's ears which would ravage his kingdom and threaten his life. Yet he thought it essential to avenge his honour, even at the risk of death.

A trap was set. Arthur went hunting for the day, leaving Lancelot behind. Agravain, Mordred and other knights hid near the queen's room. Bors warned Lancelot that there was some mischief afoot, but nothing would stop him from going to Guinevere and the lovers were caught in bed together. Lancelot fought his way out of the room. With Ector and Bors and the rest of his following, he hurried away from Camelot, intending to return and rescue Guinevere at a better opportunity. Agravain and Mordred seized the queen, taunting and insulting her while she cried piteously. When Arthur was told, he was sick at heart and sent men to arrest Lancelot, but Lancelot had gone.

Agravain and Mordred demanded that Guinevere be put to death, and it was obvious that this was what Arthur wanted. Despite protests from Gawain, the king condemned her to be burned, 'because a queen who was guilty of treachery could die in no other way, given that she was sacred'.[1] This comment suggests that the old tradition of Guinevere as a goddess had lingered on. The logic of burning a sacred queen to death seems to have been to destroy her body. Otherwise it would have to be buried and the magic force inherent in it might contaminate the earth. In the middle ages burning was the usual method of executing heretics and witches, apparently for the same reason.

The people of Camelot, who loved the queen for her kindness and courtesy, wept and railed at Arthur with impotent fury. With a strong guard, Agravain, Mordred and their brother Gaheris led Guinevere to the place of execution, the jousting-ground outside the city. Against his will, but on Arthur's orders, their youngest brother Gareth went with them. On the way he told Agravain that if Lancelot tried a rescue he would not fight him.

Lancelot and his knights were in hiding at the edge of the forest. When they heard the news of Guinevere's fate they rode hard for the jousting-ground, where the fire was already lit. Lancelot charged full tilt at Agravain, ran him through and killed him. Bors felled and killed Gaheris. Gareth, angered, turned to fight and there was a fierce mêlée between the king's men and the rescuers. Although Gareth was bare-headed, Lancelot did not recognize him and struck him a blow which split his head to the teeth. Mordred and the other survivors of his party ran away. Lancelot and his men snatched up the queen and rode for Lancelot's castle of Joyous Garde.

Arthur was plunged in grief at the deaths of his nephews. Gawain was even more appalled. When he was told what had happened, he fainted with the shock. Recovering himself, he grimly swore that he no longer wished to live except to take revenge on Lancelot for his brothers, especially for the killing of Gareth. Arthur gathered an army to besiege Lancelot in Joyous Garde.

According to Malory, these events occurred in the merry month of May:

In May, when every heart flourisheth and burgeoneth (for, as the season is lusty to behold and comfortable, so man and woman rejoiceth and gladdeth of summer coming with his fresh flowers, for winter with his rough winds and blasts causeth lusty men and women to cower and to sit by fires), so this season it befell in the month of May a great anger and unhap that stinted not till the flower of chivalry of all the world was destroyed and slain.[2]

There is now civil war in Logres and the fellowship of the Round Table is split between those who cleave to Arthur and those whose ties of family and friendship bind them to Lancelot. The disaster is not really anyone's fault, even though Agravain and Mordred act out of sheer malice. It stems from the love of Guinevere and Lancelot, a tide of passion which reason and common sense are powerless to resist. Neither of them wants to hurt Arthur, but they cannot help themselves. Nor does Lancelot wish to injure Gawain, whose friendship he values second only to that of Arthur himself. Arthur loves Guinevere and Lancelot, but he cannot accept the stain they have put on his honour. Gawain's pride of family and his feeling for his brothers drive him into an implacable vendetta against Lancelot. The killing of Gareth is the unintentional but inevitable spark which sets light to a fire that cannot be put out.

The Treachery of Mordred

Arthur and Gawain with their army set siege to Joyous Garde. Lancelot was saddened by Gareth's death and shocked to find himself in arms against Arthur. He sent a message to the king, claiming somewhat disingenuously that he had never dishonoured the queen, and offering peace. Gawain protested violently and Arthur replied that he would never make peace, for Lancelot had robbed him of those he most loved. The answer filled Lancelot with gloom, though he put a good face on it for his men's sake.

Day after day the rival knights did battle outside the castle walls and, bitter though the conflict was, neither side could withhold its admiration for the courage and skill with which its opponents fought. Arthur himself joined the fray. He rode fiercely against Lancelot, who covered himself with his shield but would take no other action against the king. Arthur knocked Lancelot off his horse. Lancelot's brother Ector galloped hastily up and dealt Arthur a blow which dazed him and then another which felled him to the ground, where he and Lancelot lay side by side. 'My lord,' said Ector to Lancelot, 'cut off his head, and our war will be over.' But Lancelot would hear none of it. He put Arthur back on his horse, mounted his own and rode away. The king returned to his army and said: 'In faith, today he has surpassed in courtesy and goodness all the knights I have ever seen; now I wish this war had never been begun, because today he has conquered my heart more with his gallantry than the whole world could have done by force.'[3]

Gawain was not to be appeased, however, and the siege went on until a message came from the Pope, ordering Arthur to take Guinevere back. Though he was angry at the Pope's interference, Arthur wanted Guinevere back and he obeyed. He sent word to the queen in the castle that if she would return to him, he and all his court would ignore anything that had been said about her relations with Lancelot.

Guinevere asked Lancelot what he wanted her to do. Lancelot insisted that she must return to Arthur. It was a question of not being shamed. He loved her more than a knight had ever loved a lady, but if she stayed with him now, her guilt and his disloyalty to Arthur would be plain for all the world to see. If she went back to her husband, face and honour could be preserved. Tears came into Lancelot's eyes as he told her this, and Guinevere began to cry. (There could scarcely be a more dramatic demonstration that shame rather than guilt was the mainspring of the code of chivalry.)

The next morning, the lovers made their sorrowful farewells. Lancelot restored Guinevere to Arthur, who ordered him to leave Britain and never set foot in the country again. Sad to leave the land which had honoured him so long, Lancelot crossed the Channel to his own estates in France. Ector, Bors, Lionel and the other knights of his following went with him. The split in the fellowship of the Round Table was not mended, for all knew that Arthur, spurred on by the relentless Gawain, would continue the war.

Arthur and Gawain mustered their forces to attack Lancelot in

France. Mordred offered to stay in England to guard the queen and, to Guinevere's anger and dismay, Arthur agreed. He put Mordred in charge of the kingdom and ordered all the people to obey him. When Arthur left for France, Guinevere wept, for her heart told her that she would never see him again.

There was heavy fighting in France when the two sides clashed, and many knights were killed. In England, meanwhile, Mordred was eaten up with desire for Guinevere. He arranged for a forged letter to be sent, ostensibly from Arthur, in which the king said that he had been mortally wounded by Lancelot, Gawain had been killed and the war was lost. The nobles in England should make Mordred king and give him Guinevere as his wife. When this letter was read out, Mordred pretended to faint with grief.

The nobles obeyed what they believed to be Arthur's dying will. When they approached Guinevere, who was sorrowing for Arthur, she said she would not marry again for she could never find so noble a husband. They told her she must be Mordred's queen, which horrified her. She knew the secret of Mordred's birth, that he was Arthur's son. The nobles insisted that Mordred was to be king and whoever was to be king she must marry. There seems to be echo here of the old tradition that Guinevere represented the land of Britain, with which the king must mate.

Guinevere took refuge in the Tower of London, where Mordred vainly besieged her. She sent a messenger in secret to France, to find out whether Arthur was alive or dead, and if he was dead, to beg Lancelot to come to her rescue.

In France the war had reached stalemate. Gawain was still bent on revenge. He challenged Lancelot to single combat, to decide the issue and prevent any more lives being lost. Lancelot was determined not to fight Gawain if he could avoid it without appearing a coward. He knew that if he had Gawain at his mercy he could not bring himself to kill him, and he was secretly worried that the fight might go against him because he had killed Gawain's brothers. He told Gawain on his oath that he had not killed Gareth knowingly. He offered to go into exile for ten years, alone as a barefoot penitent. If at the end of that time he and Gawain and Arthur were still alive, they would be reconciled to each other and the past would be forgotten.

The tears came to Arthur's eyes, for he could never have believed that Lancelot would make such an offer. He pleaded with Gawain to accept it, and so did Gawain's friend Yvain. Gawain refused. There was nothing for it but to appoint a day for the duel.

On the day, Arthur led Gawain on to the field and Lancelot was led on by Bors. The signal was given and the battle commenced. The fight between the two great paladins was the most ferocious combat ever seen between two knights. Hour after hour they thrust and hewed at each other. Their shields and helmets were hacked to pieces, their armour battered and dented, and they each suffered wounds which would have felled any lesser man. Gawain had Excalibur, Arthur's own sword, so that Lancelot was symbolically confronting the whole weight of the king's authority and prestige. But Lancelot was the younger man and, even against Excalibur, he at last gained the upper hand, when Gawain was suffering from a terrible wound in the head and was so exhausted that he could scarcely stand up or grip his sword. Lancelot begged Gawain to yield and save his life, but Gawain would not. Then Lancelot abandoned the fight and walked away.

Gawain was carried back to Arthur's camp. His wounds were treated and he was slowly recovering when a Roman army suddenly invaded Burgundy. Arthur moved against the Romans at once and defeated them. Gawain fought bravely in the battle, but the deep wound in his head re-opened and it was clear that he had not long to live.

Guinevere's messenger now arrived with the news of Mordred's treachery in England. Arthur was at first struck speechless. Then he swore to kill the traitor with his own hands, now openly admitting that Mordred was his son. The army returned to the Channel coast. The dying Gawain, carried on a horse-litter, was happy that he would die in his own country and regretted only that he could not go to Lancelot and ask his forgiveness. He urged Arthur to be reconciled with Lancelot, but the king said he did not believe that Lancelot would respond to an overture from him now.

Hearing that Arthur was crossing the Channel, Mordred raised his siege of the Tower. Guinevere took the opportunity to escape and hide herself in a nunnery. Mordred gathered a formidable army to oppose Arthur. According to Malory, he gained substantial support from the English lords: 'for then was the common voice among them that with King Arthur was never other life but war and strife, and with Sir Mordred was great joy and bliss.'[4]

Arthur landed at Dover, and there Gawain died, asking to be

buried at Camelot. His last words were: 'Jesus Christ, Father, do not judge me by my sins.'[5] Malory says that in his time Gawain's skull could be seen in a chapel at Dover Castle, with the mark of the wound Lancelot gave him, which caused his death.

Arthur has now lost his oldest and closest friends and his bravest champions. Gawain is dead, Lancelot and his kinsmen are estranged. Perceval and Galahad are gone and the Order of the Round Table has been reduced to a shadow of its former glory. Arthur's son has betrayed him and confronts him with an army stronger than his own. Catastrophe is approaching but, in the *Mort Artu*, it is not blamed on anyone. Arthur is riding the Wheel of Fortune. It has carried him up to the summit of power and fame, but the wheel turns, slowly and inexorably, and now it is taking the great king down into the depths.

The Last Act

The last scenes belong to Malory, whose account of the final tragedy is unrivalled. Moving north from Dover, Arthur found Mordred's army drawn up on Barham Down, near Canterbury. After a sharp action, Mordred drew off to the west and Arthur followed him to Salisbury Plain. Knights came to swell the king's forces, but many others joined Mordred, including those whose sympathies were with Lancelot. The quarrel with Lancelot was still exerting its poisonous effect on Arthur's fortunes.

The two armies drew up, ready for battle the next day. That night Arthur had a dream in which he saw Gawain, who warned him that if he fought Mordred he would be killed. The king should wait for Lancelot, who would come to his rescue. Arthur was impressed by the dream and in the morning he sent messengers to Mordred to patch up a truce. It was agreed that Mordred should hold Cornwall and Kent while Arthur lived, and should succeed to the throne when Arthur died. Arthur and Mordred, with fourteen knights each, were to meet between the opposing armies to ratify this agreement, but each side suspected the other of bad faith and both armies were ready to fight at the first sign of treachery. When the delegations met, it happened that a viper slithered out from under a bush and stung one of the knights in the foot. The knight drew his sword to kill it, but the atmosphere was so tense with suspicion that as soon as the hostile armies saw a sword drawn, they charged into battle. 'And never since was there a more dolefuller battle in a Christian land, for there was but rushing and riding, foining [thrusting] and striking, and many a grim word was there spoken of either to other, and many a deadly stroke.'[6]

The battle raged all day and the slaughter was terrible. Arthur and Mordred both fought like lions. The knights of the Round Table were cut down until only two were left alive, Bedivere and Lucan, who was severely wounded. 'Jesu mercy!' said the king, 'where are all my noble knights become? Alas, that ever I should see this doleful day. For now,' said King Arthur, 'I am come to mine end.'

The two sides had fought each other to a standstill. Arthur saw Mordred leaning on his sword among a pile of corpses. Determined to make an end of the traitor, the king called for his spear. Lucan tried to dissuade him, telling him that the victory was won and reminding Arthur of his dream and the prediction that he would be killed. 'Now tide me death, tide me life,' said the king, 'now I see him yonder alone, he shall never escape mine hands. For at a better avail shall I never have him.'

Then Arthur took his spear in both hands and ran at Mordred, shouting, 'Traitor, now is thy death-day come!' The spear struck Mordred in the body, below his shield, and ran him through. 'And when Sir Mordred felt that he had his death's wound he thrust himself with the might that he had up to the burr [hand-guard] of King Arthur's spear, and right so he smote his father, King Arthur, with his sword holding in both his hands, upon the side of the head, that the sword pierced the helmet and the tay [membrane] of the brain. And therewith Mordred dashed down stark dead to the earth.'

Despite the prediction of his own death, Arthur had killed Mordred with his own hands, as he had sworn to do. True to himself at the last, he valued his honour above his life. Bedivere and Lucan carried the fainting king away from the battlefield, where looters were pillaging the dead. The strain was too much for the badly wounded Lucan, and he collapsed and died. Arthur and Bedivere were left alone. Arthur ordered Bedivere to take his sword, Excalibur, and throw it into a lake which was nearby. Bedivere took the sword, but on his way to the water he thought it was a pity to throw away so fine a weapon. He hid it by a tree and went back to Arthur. The king asked him what he had seen, and Bedivere said he had seen nothing. Arthur told him he had not carried out his orders, and to go and do so. Bedivere again tried to deceive Arthur, but when the king sent him back a third time he hurled the great sword as far out into the lake as he could. 'And there came an arm and an hand, and took it and cleight [seized] it, and shook it thrice and brandished, and then vanished with the sword into the water.'

The otherworld sword, the symbol of Arthur's manhood and kingship, had returned to the home from which it came, and it was time for the king himself to leave the human world. Bedivere hoisted Arthur on his back and carried him to the water's edge. By the bank was a barge with many fair ladies in it, all wearing black hoods. Among them were Morgan le Fay and the Lady of the Lake. Bedivere gently lowered Arthur into the barge, where Morgan and the ladies received him sorrowfully. They rowed away from the land and Bedivere, left alone on the bank, cried out: 'Ah, my lord Arthur, what shall become of me, now you go from me and leave me here alone among mine enemies?' 'Comfort thyself,' said the king, 'and do as well as thou mayest, for in me is no trust for to trust in. For I must into the vale of Avilion to heal me of my grievous wound. And if thou hear nevermore of me, pray for my soul.'

Weeping, Bedivere took to the forest and the next day came to Glastonbury, where he found a chapel and a hermitage. The hermit told him that a company of ladies had brought the dead body of King Arthur there at midnight and he had buried the king in the chapel. Bedivere stayed at the hermitage, vowing to spend the rest of his life there in prayer for Arthur.

Guinevere was hiding in a nunnery at Amesbury, on the edge of Salisbury Plain. When she heard that both Arthur and Mordred were dead, she took her vows and became a nun. Lancelot and his kinsmen, meanwhile, were hurrying to England to fight for Arthur against Mordred. They arrived at Dover to discover that they were too late. Lancelot went to Gawain's tomb, where he wept and prayed for Gawain's soul. Then he rode alone to Amesbury.

When Guinevere saw Lancelot she fainted. As soon as she could speak, she told Lancelot that it was because of their love that Arthur and the noblest knights in the world had been slain. She begged Lancelot never to see her again, for she intended to devote her remaining days to religion and to pray that God would forgive her misdeeds. Lancelot should go back to his own country, find himself a wife and live with her happily.

Lancelot said he could not be so false and he would never take a wife. He had hoped that Guinevere would return to France with him, but now he too meant to forsake the world. He would find a hermitage and spend his life in prayer. He asked Guinevere to give him a last kiss. '"Nay," said the queen, "that shall I never do, but abstain you from such works." And they departed; but there was never so hard an hearted man but he would have wept to see the dolour that they made, for there was lamentation as they had been stung with spears, and many times they swooned. And the ladies bore the queen into her chamber.'[7]

Sadly Lancelot rode away to Glastonbury, where he found Bedivere at the hermitage. He settled down there and served God with prayer and fasting. Presently Bors arrived, and later seven other knights. They all stayed on at the hermitage, remote from the world. Six years passed by, and Lancelot was ordained priest. A year after this, word came that Guinevere had died at Amesbury. Lancelot and the others went to the nunnery and took the queen's body back to Glastonbury on a horse-drawn bier. They buried her in the chapel beside Arthur.

From then on Lancelot scarcely ate or drank anything and after a time it was clear that he was dying. He asked to be buried at Joyous Garde. One night, one of his companions dreamed that he saw angels carrying Lancelot up into heaven and the gates of heaven opening for him. When he told the others this, they hurried to Lancelot's cell, where his body was lying lifeless on the bed. He was smiling.

They took Lancelot to Joyous Garde on the same bier that had carried Guinevere's body to Glastonbury. There in his own castle, as the custom was, he lay in an open coffin so that all the people might seen him. His brother, Ector de Maris, came to Joyous Garde, strode into the hall and looked down at the body in the coffin.

'Ah, Lancelot,' he said, 'thou were head of all Christian knights. And now I dare say,' said Sir Ector, 'thou Sir Lancelot, there thou liest, that thou were never matched of earthly knight's hand. And thou were the courteousest knight that ever bore shield. And thou were the truest friend to thy lover that ever bestrode horse, and thou were the truest lover, of a sinful man, that ever loved woman, and thou were the kindest man that ever struck with sword. And thou were the goodliest person that ever came among press of knights, and thou was the meekest man and the gentlest that ever ate in hall among ladies, and thou were the sternest knight to thy mortal foe that ever put spear in the rest.'[8]

Lancelot was buried at Joyous Garde. Bedivere returned to the hermitage at Glastonbury, where he remained till his death. Bors and Ector went crusading to the Holy Land, where they fought valiantly against the Saracens, and there they both breathed their last.

Arthur had been interred at Glastonbury, or so it was supposed, but the hermit there could not be quite certain that it was Arthur's body which he had laid to rest in the chapel.

> Yet some men say in many parts of England that King Arthur is not dead, but had by the will of our Lord Jesu into another place; and men say that he shall come again and shall win the Holy Cross. Yet I will not say that it shall be so, but rather I would say: here in this world he changed his life. And many men say that there is written upon the tomb this: HIC IACET ARTHURUS REX QUONDAM REXQUE FUTURUS.[9]

'Here lies Arthur, king once and king to be.' The belief that Arthur would return lingered on long after Malory's time. In Welsh folk tradition his knights are sleeping in a cave on Snowdon, but the king himself lies under a cairn on the pass of Bwlch y Snethau. At Cadbury Castle in Somerset, Arthur and his men sleep in a cavern under the hill and sometimes at night the beat of horses' hooves can be heard as they ride down to a spring beside a nearby church. In the north of England the king is asleep with Guinevere and his court and his hunting dogs around him, deep beneath Sewingshields Castle in Northumberland. There is a horn on the table beside him and when it is sounded, the great hero will wake. Arthur represented something too noble and glorious to be relegated entirely to the past with no hope of recovery. In his hold on human hearts he remained 'the once and future king'.

The truth of this was demonstrated soon after Caxton published *Le Morte Darthur* in July 1485. The end of the Wars of the Roses and the era of violence in which Malory had lived was now almost in sight. Within a few weeks Henry Tudor launched his long-threatened invasion from France, won the battle of Bosworth and seized the crown of England, which had now changed hands six times in a quarter of a century. For its security and the restoration of peace and order the new dynasty needed to make a deeper appeal to its subjects' allegiance than the argument of force The appeal was found in the legend of Arthur. On his father's side Henry VII was Welsh. He fought under the standard of the Red Dragon, the symbol of the Britons in Geoffrey of Monmouth, and he claimed the throne in right of a genealogy which traced his ancestry back to Arthur. When his son and heir was born he named him Arthur and he made sure that the child was born and christened at Winchester, which Malory had identified as Camelot. The implication was that the House of Tudor under King Arthur II would recreate Britain's heroic golden age. As it proved, Arthur II never reigned. He died young and Henry VII was succeeded by his second son, Henry VIII. Undeterred, the antiquary John Leland hailed Henry VIII as Arthur Restored.

Malory and his predecessors had created a powerful and enduring myth which has exercised a compelling attraction over writers and artists ever since. In Spenser's *Faerie Queen* the adventures of the knights serve the purposes of Tudor propaganda and the young Arthur is in love with Gloriana, the Fairy Queen, who is an idealized portrait of Queen Elizabeth. Milton seriously contemplated writing an Arthurian epic, though in the end his republican principles would not permit it, but Dryden wrote a successful opera, *King Arthur*, with music by Henry Purcell. Though temporarily out of favour in the Age of Reason, in the nineteenth century the Matter of Britain cast its spell on two of the greatest creative talents ever drawn to it, Wagner and Tennyson. Wagner became interested in Wolfram von Eschenbach's *Parzival* in the early 1840s, when he was planning operas on two other legendary heroes, Tannhäuser and Lohengrin. He also read Gottfried von Strassburg, and his *Tristan und Isolde* was first produced in 1865. His last and perhaps finest work, the Grail opera *Parsifal*, was staged in 1882, a few months before the composer's death.

By this time in England, Tennyson had almost completed his magnificent and immensely popular *Idylls of the King*, which occupied him intermittently for most of his life. In the dedication, to Queen Victoria, he said that his version of the old tale was:

> New-old, and shadowing Sense at war with Soul,
> Ideal manhood closed in real man,
> Rather than that gray king, whose name, a ghost,
> Streams like a cloud, man-shaped, from mountain peak,
> And cleaves to cairn and cromlech still.

From Tennyson the Arthurian torch passed to the Pre-Raphaelites, Swinburne, William Morris and Hardy. In our own century the Matter of Britain has attracted poets and historical novelists who include

John Masefield, Charles Williams, T. H. White, Rosemary Sutcliff and Mary Stewart. Outside the Arthurian field itself, the hero who risks his life for his ideal of what he ought to be, the man who takes the law into his own hands to right wrongs, is the central character of innumerable novels, thrillers and Westerns in print and on the screen.

'Bravery,' it has been said, 'never goes out of fashion.' The tales of Arthur and his knights have continued to impress and inspire long after the disappearance of the world in which they are set. Their setting, in a deeper sense, is not the middle ages but the world in which each of us makes the journey from birth to death. Their jousts and combats and valorous adventures are metaphors for all the struggles and battles of life. Their values of courage, loyalty and honour are not cheapened by circumstances or tarnished by time. Their themes are always relevant: the need for action, risk and danger, the conflict between action and ease, and between action and love, the hope of immortality and the longing to score a defiant victory against death, the last enemy. Above all perhaps, in our age of uncertainty, the central Arthurian theme has a particular urgency and appeal. This is the search for integrity, the attempt to find and realize one's true and best self: 'Ideal manhood closed in real man.'

Appendix I

The Matter of Britain

The following notes on authors, books and stories mentioned in the text are in alphabetical order. Where no author's name is given, the author is unknown.

Aneirin, *The Gododdin*: in Welsh verse, *c.* 600; if the reference to Arthur is not an interpolation, it is the earliest on record.

El Baladro del Sabio Merlin (The Cry of the Wise Merlin): Spanish rendering of the *Suite du Merlin*, fifteenth century.

Beroul, Norman poet: author of a romance of Tristan in French, *c.* 1190; only part of it has survived.

The Black Book of Carmarthen: a collection of early Welsh poems, compiled *c.* 1200.

The Book of Taliesin: a collection of early Welsh poems, compiled *c.* 1275. Taliesin was a famous bard of the sixth century, and some of the poems may be his.

'Branwen Daughter of Llyr': a story in the Welsh *Mabinogion*, thought to have been written down in the eleventh century. The hero, Bran, was probably the prototype of the Fisher King in the Grail legends.

Cambrian Annals: see Welsh Annals.

Caradoc of Llancarfan, *Vita Gildae* (Life of Gildas), *c.* 1130: includes the story of Guinevere's abduction and also says that Glastonbury was called the Isle of Glass.

Chretien de Troyes: French poet, a leading architect of the literary genre of Arthurian romance. He lived at the court of the Counts of Champagne at Troyes in eastern France, where his patroness was the Countess Marie, daughter of Eleanor of Aquitaine by her first husband, King Louis VII of France. He is the first writer whose work has survived to mention Camelot, Lancelot and the love affair between Lancelot and Guinevere. His Arthurian stories deal with the adventures of knights of Arthur's court, with Arthur himself largely relegated to the background. The poems are based on Celtic tales which were already in circulation, but this does not mean that Chrétien was merely a passive retailer of other men's stories or that his only creative nourishment was Celtic. He was well read in classical poetry and he brought his own considerable equipment of psychological insight, sophisticated irony and powers of organization to bear on his material. His first surviving poem, for example, *Erec et Enide*, he took from a tale told by travelling minstrels, which he said he had turned into *une molt bele conjointure*, which seems to mean 'a beautiful construction'.

Chrétien's tales of the Round Table are usually dated roughly between 1160 and 1180. They are:

1 *Erec et Enide*
2 *Cligès*
3 *Le Chevalier de la Charrette* (The Knight of the Cart) or *Lancelot*: suggested to the author by the Countess Marie.
4 *Le Chevalier au Lion* (The Knight with the Lion) or *Yvain.*
5 *Le Conte du Graal* (The Story of the Grail) or *Perceval*: this is the earliest surviving story about the Grail. Chrétien said he based it on a book given him by Count Philip of Flanders, so whatever he made of the Grail he apparently did not invent it. The story is unfinished.

After Chrétien's death, four different Continuations, or sequels, to the *Conte du Graal* appeared, all in French verse:

1 First Continuation, *c.* 1200, includes Gawain's visit to the Grail castle.

2 Second Continuation, *c.* 1200, attributed to Wauchier de Denain, continues Perceval's adventures.
3 Manessier's Continuation, *c.* 1210–20, tells how Perceval won the Grail.
4 Gerbert's Continuation, *c.* 1230, an alternative conclusion to Perceval's adventures.

Le Conte du Graal: see Chrétien de Troyes.

'Culhwch and Olwen': a story in the Welsh *Mabinogion*, believed to have reached its present form by 1100.

Didot *Perceval*: named for a Parisian bookseller who once owned one of the manuscripts. In French prose, early thirteenth century, a continuation of Robert de Boron's *Joseph* and *Merlin*.

Draco Normannicus by Etienne de Rouen, *c.* 1170: sardonic treatment of Arthur's return.

Eilhart von Oberge: German poet, author of *Tristan*, *c.* 1170.

L'Estoire del Saint Graal: see Vulgate Cycle.

First Continuation: see Chrétien de Troyes.

Geoffrey of Monmouth: probably of Welsh or Breton descent, he presumably came from Monmouth on the Welsh–English border. His father's name was Arthur, which may imply a family tradition of interest in the hero. He made his way in the Church and seems to have written his books while teaching at Oxford, which was already a scholarly centre, though the university was not yet in existence. He was given a Welsh bishopric in 1152 and died in 1155. His interest in Celtic traditions is evident from his books:

1 *Prophetiae Merlini* (The Prophecies of Merlin), which he soon afterwards included in
2 *Historia Regum Britanniae* (History of the Kings of Britain), *c.* 1136, containing the first full and connected account of Arthur that has survived.

3 *Vita Merlini* (Life of Merlin), in verse, *c.* 1148: based on Welsh lore which Geoffrey apparently did not know when he wrote his *History*.

Of these, much the most important is the *History*, which has been described as the most successful historical novel ever written. It was extremely influential because it was generally accepted for four centuries as an authentic history of Britain from the earliest times to the Saxon conquest. Geoffrey said that he based it on 'a very old book in the British language', which had been given him by a friend. The British language could mean Welsh or Breton, but it is very doubtful whether the book ever existed.

It used to be the fashion to look disapprovingly down one's nose at Geoffrey of Monmouth as an unprincipled liar who made his history up out of his own head and passed it off as genuine. However, besides using such written sources as Gildas and Nennius, he almost certainly relied on oral traditions. He himself says that these were plentiful and he has probably preserved many traditional Welsh, Cornish and Breton tales. It is still true that the bulk of the *History* is fiction (and enjoyable fiction at that), but Geoffrey himself may not have made up quite as much of it as was once supposed.

Gerald of Wales (Giraldus Cambrensis): Norman historian and civil servant, born in Wales *c.* 1146. He described the discovery of Arthur's grave at Glastonbury in *De Principis Instructione*. He died *c.* 1223.

Gildas: British monk, author of *De Excidio Britanniae* (The Ruin of Britain), *c.* 540, which deals with the British war against the Saxons, but does not mention Arthur.

The Gododdin: see Aneirin.

Gottfried von Strassburg, German poet, author of *Tristan*, *c.* 1210, generally considered the finest medieval treatment of the legend, though it was soon superseded in popular favour by the Prose Tristan: based on a romance in French verse, written in England *c.* 1160 by a poet named Thomas, only a fragment of which has survived. The last part of Gottfried's poem is also lost.

Hartman von Aue: German poet, author of *Erek*, *c.* 1190, based partly on Chrétien de Troyes' *Erec*, and of *Iwein*, *c.* 1200, a version of Chrétien's *Yvain*.

'The Lady of the Fountain': a story in the Welsh *Mabinogion*. The hero is Owain son of Urien, who in the French romances is Yvain: the date and relationship to Chrétien de Troyes' *Yvain* are disputed.

Layamon: English priest who lived at Arley Regis in Worcestershire, author of the *Brut*, *c.* 1190, an expanded rendering in English verse of Wace's *Brut*: the first surviving evidence of the native English adopting Arthur as a hero.

The Mabinogion, a collection of eleven Welsh stories of varying date, preserved in the White Book of Rhydderch, *c.* 1325, and the Red Book of Hergest, *c.* 1400. The first four stories, or 'branches', contain almost all that is left of the legend of Pryderi, who was probably the prototype of Perceval.

Malory, Sir Thomas: born *c.* 1410 at Newbold Revell, a village in Warwickshire, he inherited his family estates in 1434. In 1450 he embarked on a career of violent crime. With a gang of twenty-six men he attempted to ambush and murder the Duke of Buckingham. He twice broke into the house of a woman named Joan Smyth and raped her. He carried off sheep and cattle. He extorted money from various people with menaces. At the head of a small army of a hundred men he smashed down the doors of a monastery with battering-rams, terrorized the monks and carried off money and valuables.

There is no record of Malory being tried for any of these offences, but he was held in a succession of prisons, from which he twice escaped and twice was released on bail. It was a singularly violent period, in which England was torn by an intermittent civil war. Not only did the rival armies of York and Lancaster tramp about the country, plundering and destroying, but miscellaneous bands of armed ruffians took advantage of the weakness of central government to kill, rape and loot. Malory was evidently one of these ruffians.

In 1462 Malory served with the Earl of Warwick on the Yorkist side in the civil war and he seems to have followed Warwick over to the Lancastrian side later. It may have been for this political misjudgement that in 1469 he was imprisoned in Newgate jail in London. It was there that he wrote the last words of 'the whole book of King Arthur and his noble knights of the Round Table'. He died in 1471 and was buried in Greyfriars Church, across the road from the prison.

That a book of such nobility should have been written by a violent criminal surprises even those who do not share the common delusion that a great artist must also be a good man. It has been suggested that Malory was not guilty of the accusations against him, or even, desperately, that the author must have been someone else of the same name. But it is not really so astonishing that a man of fiercely aggressive temperament should feel the attraction of the heroic ideal. Malory differs from many other romantic writers, not in having an appetite for violence, but in expressing it in his actions as well as on paper.

In 1485 Malory's book was edited and printed by William Caxton, who gave it the title *Le Morte Darthur*. Caxton's version was the only one known until 1934, when an earlier copy of Malory was discovered at Winchester College. Professor Eugene Vinaver, who published a masterly edition of the manuscript, came to the conclusion that Malory had not set out to write a single unified book. He wrote eight separate books, each of which picked out one strand from the tapestry which the French romances had woven. This conclusion, however, has been much disputed.

Malory's principal sources were the Vulgate Cycle, the Prose Tristan and the *Suite du Merlin*. His handling of his sources is sometimes confused and inconsistent, but his work is the finest monument ever raised to the heroic ideals of chivalry.

Manessier's Continuation: see Chrétien de Troyes.

La Mort Artu: see Vulgate Cycle.

Nennius: Welsh monk, author of *Historia Brittonum* (History of the Britons), *c.* 800, the earliest surviving work to identify Arthur as the British war-leader who defeated the Saxons at Badon.

'Peredur Son of Efrawg': a story in the Welsh *Mabinogion*, dated to the early thirteenth century. The hero, Peredur, is Perceval, but the story's relationship to Chrétien de Troyes' *Conte du Graal* is in doubt.

Perlesvaus, in full *Le Haut Livre du Graal: Perlesvaus* (The High Book of the Grail: Perlesvaus): in French prose, possibly *c.* 1210, but the date is very uncertain. It deals with the adventures of Perceval (Perlesvaus), Lancelot and Gawain in quest of the Grail. The author described it as a translation from a Latin book in the possession of Glastonbury Abbey, but whether this was true is doubtful.

Post-Vulgate Romance: modern name for a work in French prose, *c.* 1230–40, by an author pretending to be Robert de Boron. Not all of this attempt to put together a complete account of the Matter of Britain has survived. Part which has is the *Suite du Merlin* (Sequel of Merlin), which is a continuation of the Vulgate *Merlin.*

Prose Lancelot: see Vulgate Cycle.

Prose Tristan: modern name for a romance in French, *Le Roman de Tristan de Léonois*, which turned Tristan into a knight of the Round Table. The first version of it was written *c.* 1230, perhaps in England: a second, expanded version dates from *c.* 1250.

La Queste del Saint Graal: see Vulgate Cycle.

Renaud de Beaujeu: French poet, author of *Le Bel Inconnu* (The Fair Unknown), *c.* 1185–90, the earliest surviving version of the Fair Unknown story.

Robert de Boron: writing *c.* 1200. He presumably came from the village of Boron near Montbéliard in Burgundy; author of two poems, now known as *Joseph d'Arimathie* and *Merlin*. The *Joseph* is the earliest surviving account of the early history of the Grail. Robert said he drew it from a 'great book', written by 'great clerks'. Only a fragment of the *Merlin* has survived, but a prose version of it is preserved in the Vulgate *Merlin*.

Sir Gawain and the Green Knight: a masterpiece by an English poet, late fourteenth century.

La Suite du Merlin: see Post-Vulgate Romance.

Ulrich von Zatzikhoven: Swiss priest, author of *Lanzelet*, a version of Lancelot's adventures in German verse, *c.* 1200. Ulrich said he translated it from a book in French, apparently written in England, given him by Hugh de Morville, a courtier of King Richard I.

Vulgate Cycle: modern name for the first and most influential complete account of the Matter of Britain, in French prose, *c.* 1215–30, so called because of its semi-canonical standing as the 'bible' of Arthurian literature. It has five branches or sections:

1 *L'Estoire del Saint Graal* (The History of the Holy Grail), the early history of the Grail and how it was brought to Britain.
2 *Merlin*: the story of Merlin and Arthur's early career, consisting of a prose version of Robert de Boron's *Merlin* plus a continuation.
3 *Lancelot*: the adventures of Lancelot and other knights of the Round Table, introducing a new character, Galahad.
4 *La Queste del Saint Graal* (The Quest of the Holy Grail): how the knights went in quest of the Grail and how Galahad was successful; permeated by Cistercian ideals.
5 *La Mort le Roi Artu* (The Death of King Arthur): the final tragedy.

This is the logical order of the branches, but not the order in which they were written. The last three were written first. They are grouped together and called the Prose Lancelot, because Lancelot is the leading character almost throughout. Different authors seem to have written them, but it is thought that the Prose Lancelot must have been planned by one person. The *Estoire* and the *Merlin* were added later, to complete the whole chronicle.

No one knows who the authors of the Vulgate Cycle were, but it is generally agreed that the Prose Lancelot's claim that it was written by Walter Map at the request of King Henry II of England is false. Walter Map was a Norman-Welsh literary man and courtier, who died in 1209. The attribution to him at least suggests that Henry II was known to have taken a keen interest in the Matter of Britain.

Wace, Robert: author of *Roman de Brut* (Story of Brutus), 1155, an enlarged version in French verse of Geoffrey of Monmouth's *History*. Wace came from Jersey and spent most of his life in Normandy. He is the first author to mention the Round Table and he says that the Bretons told many tales of it.

Welsh Annals (*Annales Cambriae*): Welsh historical records, compiled *c.* 950, the earliest surviving source to mention the deaths of Arthur and Mordred in battle.

William of Malmesbury: Norman historian, born *c.* 1090, librarian at Malmesbury Abbey. His *De Gestis Regum Anglorum* (Chronicle of the English Kings) came out *c.* 1140; he also wrote a history of Glastonbury, *De Antiquitate Glastoniensis Ecclesiae.*

Wolfram von Eschenbach: German poet, author of *Parzival*, *c.* 1210. This story of Perceval's Grail quest is based on Chrétien de Troyes' *Conte du Graal*, but there are puzzling references to a Provencal poet named Kyot, said to have found the story of the Grail written in Arabic, at Toledo in Spain, and to have passed it on to Wolfram. Whether any such person existed is very much in doubt.

Appendix II
Miss Weston and A. E. Waite

Jessie L. Weston's theory about the Grail has not worn well with scholars, but it has influenced many other writers and has become part of modern occult lore. It seems to have sprung from or been partly inspired by occult lore to begin with. Miss Weston was closely in touch with a small occult group called the Quest Society, and her books leave no doubt about her sympathy with its outlook. The Quest Society was founded in London in 1909 by G. R. S. Mead, who published translations of gnostic and Hermetic Graeco-Egyptian texts. Earlier in his career he had been secretary to Madame Blavatsky, the formidable leader of the Theosophists.

Like other groups of its kind, the Quest Society believed in a 'secret tradition' of wisdom, stemming from the classical mystery religions and the high magic of the early Christian world, and concerned with the means by which man could become divine. The secret tradition was supposedly handed down by word of mouth for centuries, from generation to generation of the initiated. Miss Weston appears to have been influenced by this belief and from her point of view it removed one of the principal objections to her theory, the lack of evidence for the existence of a secret cult which inspired the Grail legends. Occultists also believed that the secret wisdom had been known to the Knights Templar and that the Templars were connected with the inner mystery of the Grail.

Another believer in the secret tradition of wisdom was A. E. Waite, the author of numerous massive books on mysticism and the occult. Waite pointed out an apparent parallel between the four 'hallows' or sacred objects in the Grail castle and the four suits of the mysterious Tarot pack.

Grail Hallows	*Tarot Suits*
Cup	Cups
Lance	Wands
Dish	Coins
Sword	Swords

Miss Weston, who was in touch with Waite, put this parallel forward in her book *From Ritual to Romance*: to the marked irritation of Waite, who made sulphurous comments about her brand of occultism in his own book *The Holy Grail*.

The parallel is often mentioned by writers on the Tarot, but its legs are remarkably wobbly. The cup and the sword are common to both sets, but it is highly questionable whether the Tarot wand is really a lance or the coin a dish. In any case, there are not four sacred objects in the Grail castle. There may be less than four or more than four. There may or may not be a dish and the sword, though often present, is usually not one of the hallows at all.

In the *Conte du Graal* the sword is brought in and presented to Perceval. It is a special sword, certainly, but there is no suggestion that it is sacred. The Grail procession then enters, with the lance followed by the Grail itself and the carving-dish. In *Perlesvaus* Gawain sees the Grail and the lance in procession, but nothing else. In *Parzival* the lance is carried across the hall first by a young man. Then come two candlesticks, four candles and a table, four more candles and two silver knives, six glass lights and the Grail itself, borne in by a total of twenty-five beautiful women: but no dish and no sword. The whole notion of a set of four sacred objects in the Grail castle seems to be illusory.

Notes

The following abbreviations are used in these notes:

ALMA – *Arthurian Literature in the Middle Ages*, edited by R. S. Loomis.

Malory C. – Malory's *Le Morte Darthur*: the Caxton version, edited by A. W. Pollard; references are to books and chapters.

Malory W. – The Winchester manuscript of Malory, edited by Eugene Vinaver; references are to page numbers.

Full details of all works mentioned below appear in the Bibliography on p. 214.

Chapter 1: *Arthur of Britain*

1 Quoted in Morris, *Age of Arthur*, pp. 57, 75.
2 *ALMA*, p. 3.
3 Chambers, *Arthur of Britain*, pp. 1–2.
4 William of Malmesbury, *Chronicle*, 1.1.
5 Jackson, *Gododdin*, p. 101.
6 *Mabinogion*, Jones edn, p. 98.
7 Jackson, *Gododdin*, p. 107.
8 Ibid., pp. 139, 140.
9 *Tain*, p. 184.
10 Quoted in Ross, *Everyday Life*, p. 214.
11 *Mabinogion*, Jones edn, p. xxiv.
12 *ALMA*, p. 13.
13 This explanation is given in *Perlesvaus*; see *High History of the Holy Grail*, p. 244.
14 Rees, *Celtic Heritage*, p. 224.
15 Geoffrey of Monmouth, *History*, 11.2.
16 Wace and Layamon, *Arthurian Chronicles*, pp. 114, 264.
17 Chambers, *Arthur of Britain*, pp. 112–13.

Chapter 2: *Arthur and the Round Table*

1 Malory W. 12; C. 1.5.
2 Malory W. 15; C. 1.6.

3 *Quest of the Holy Grail*, p. 99.
4 Chrétien de Troyes, *Arthurian Romances*, p. 184.
5 *High History of the Holy Grail*, p. 26.
6 Malory W. 372; C. 8.1.
7 Malory W. 718–19; C. 10.64.
8 Malory W. 269; C. 6.10.
9 Gottfried von Strassburg, *Tristan*, pp. 214–18: there is a similar episode in the Prose Tristan, followed by Malory.
10 Chrétien de Troyes, *Arthurian Romances*, p. 247.
11 Malory W. 1203–4; C. 20.17.
12 Chrétien de Troyes, *Arthurian Romances*, p. 295.
13 Malory W. 744; C. 10.73.
14 *High History of the Holy Grail*, p. 133.
15 For this story and what follows, see Malory W. 791 ff.; C. 11.1 ff.
16 Malory W. 817; C. 12.1.
17 Malory W. 830; C. 12.8.
18 Malory W. 1151–2; C. 19.12.
19 Chrétien de Troyes, *Arthurian Romances*, p. 211.
20 Wolfram von Eschenbach, *Parzival*, section 508.
21 Gottfried von Strassburg, *Tristan*, p. 206: cf. *ALMA*, pp. 154–5.
22 *ALMA*, p. 21; *Guide to Welsh Literature*, p. 104; for Merlin's early history, see also Jarman, *Legend of Merlin*.

Chapter 3: *The Quest of the Grail*

1 Mark 14:24.
2 Malory W. 865; C. 13.7: Malory's source is the *Queste del Saint Graal*, see *Quest*, pp. 43–4.
3 *ALMA*, p. 274.
4 Chrétien de Troyes, *Perceval*, lines 3224–5.
5 John 8:12.
6 See Olschki, *Grail Castle*.
7 Loomis, *Grail*, pp. 72–3.
8 John 19:34.
9 Gerald of Wales, *Topography of Ireland*, 3.25: cf. Ross, *Everyday Life*, p. 122.
10 Matthew 7:7.
11 Robert de Boron, lines 932–6.
12 See *Acts of Pilate* in *Apocryphal New Testament*.
13 Robert de Boron, lines 3332–6
14 Numbers 16:31.
15 *Mabinogion*, Jones edn, p. 29.
16 Loomis, *Grail*, pp. 60–1, and in *ALMA*, pp. 287–8.
17 Revelation 17:4.
18 Wolfram von Eschenbach, *Parzival*, sections, 235 ff.
19 Mark 12:10.
20 Wolfram von Eschenbach, *Parzival*, section 484.

21 Genesis 31:47–8.
22 *Quest*, p. 276.
23 Ibid., p. 283.
24 1 Corinthians 2:9.

Chapter 4: *The Passing of Arthur*

1 *Death of Arthur*, p. 120.
2 Malory W. 1161; C. 20.1.
3 *Death of Arthur*, p. 145.
4 Malory W. 1228; C. 21.1.
5 *Death of Arthur*, p. 200.
6 For this and the following quotations, see Malory W. 1235 ff.; C. 21.4–5.
7 Malory W. 1253; C. 21.10.
8 Malory W. 1259; C. 21.13.
9 Malory W. 1242; C. 21.7.

Bibliography

Alcock, Leslie, *Arthur's Britain* (Allen Lane, London, 1971).

The Apocryphal New Testament, ed. M. R. James (Clarendon Press, Oxford, 1972 reprint).

Arthurian Literature in the Middle Ages, ed. R. S. Loomis (Clarendon Press, Oxford, 1959).

Ashe, Geoffrey, *King Arthur's Avalon* (Collins, London, 1957).

——, *Camelot and the Vision of Albion* (Heinemann, London, 1971).

—— (ed.), *The Quest for Arthur's Britain* (Pall Mall Press, London, 1968).

Barber, Richard, *Arthur of Albion* (Barrie & Rockliff, London, 1961).

——, *The Knight and Chivalry* (Longman, London, 1970).

——, *King Arthur in Legend and History* (Boydell Press, Ipswich, 1973).

Bennett, J. A. W. (ed.), *Essays on Malory* (Clarendon Press, Oxford, 1963).

Beroul, *The Romance of Tristan*, trans. A. S. Fedrick (Penguin Books, 1970, paperback).

Bogdanow, Fanni, *The Romance of the Grail* (Manchester University Press and Barnes & Noble, N.Y., 1966).

Bowra, C. M., *Heroic Poetry* (Macmillan, London, and St Martin's Press, N.Y., 1966).

Campbell, Joseph, *The Hero with a Thousand Faces* (Princeton University Press, 1972).

Chadwick, N. K., *Celtic Britain* (Thames & Hudson, London, 1963).

Chambers, E. K., *Arthur of Britain* (Speculum Historiale, Cambridge, and Barnes & Noble, N.Y., 1964 reprint).

Chrétien de Troyes, *Arthurian Romances*, trans. W. W. Comfort (Dent, London, and Dutton, N.Y., Everyman's Library, 1958 reprint).

Chrétien de Troyes, *Le Roman de Perceval ou Le Conte du Graal*, ed. W. Roach (Librairie Droz, Geneva, 2nd edn, 1959).

The Death of King Arthur, trans. J. Cable (Penguin Books, 1971, paperback) (the Vulgate *Mort Artu*).

Didot *Perceval*, see *Romance of Perceval*.

Dillon, M. and Chadwick, N. K., *The Celtic Realms* (Weidenfeld & Nicolson, London, 1967).

Geoffrey of Monmouth, *The History of the Kings of Britain*, ed. L. Thorpe (Penguin Books, 1966, paperback).

Gerald of Wales (Giraldus Cambrensis), *Historical Works* (Bohn, London, 1863).

Gottfried von Strassburg, *Tristan*, trans. A. T. Hatto (Penguin Books, 1960, paperback).

Grinsell, L. V., 'The Legendary History and Folklore of Stonehenge', *Folklore*, volume 87 (1976).

Guide to Welsh Literature, ed. A. O. H. Jarman and G. R. Hughes (Christopher Davies, Swansea, volume 1 1976).

The High History of the Holy Grail, trans. S. Evans (James Clarke, Cambridge, 1969 reprint) (a translation of *Perlesvaus*).

Huizinga, J., *The Waning of the Middle Ages* (Penguin Books, 1965, paperback reprint).

Jackson, K. H., *The Gododdin* (Edinburgh University Press, 1969).

Jarman, A. O. H., *The Legend of Merlin* (University of Wales Press, Cardiff, 1976, paperback reprint).

Jung, E. and von Franz, M. L., *The Grail Legend*, trans. A. Dykes (Hodder & Stoughton, London, 1971).

Kelly, T. E., *Le Haut Livre du Graal, Perlesvaus* (Librairie Droz, Geneva, 1974).

Layamon, see Wace and Layamon.

Loomis, R. S., *Wales and the Arthurian Legend* (University of Wales Press, Cardiff, 1956).

——, *The Grail: From Celtic Myth to Christian Symbol* (University of Wales Press, Cardiff, and Columbia University Press, N.Y., 1963).

——, *The Development of Arthurian Romance* (Hutchinson, London, 1963).

Luttrell, C., *The Creation of the First Arthurian Romance* (Edward Arnold, London, 1974).

The Mabinogion, ed. G. and T. Jones (Dent, London, and Dutton, N.Y., Everyman's Library, 1949).

The Mabinogion, trans. J. Gantz (Penguin Books, 1976, paperback).
Malory, Sir Thomas, *Le Morte Darthur*, ed. A. W. Pollard (Medici Society, London, 1929) (the Caxton version).
——, *Works*, ed. E. Vinaver (Clarendon Press, Oxford, 2nd edn, 1973 reprint, three volumes) (the Winchester manuscript).
Medieval Romances, ed. R. S. and L. H. Loomis (Modern Library, N.Y., 1957, paperback).
Morris, J., *The Age of Arthur* (Weidenfeld & Nicolson, London, 1973).
Olschki, L., *The Grail Castle and Its Mysteries*, trans. J. A. Scott (Manchester University Press, 1966).
Owen, D. D. R., *The Evolution of the Grail Legend* (Oliver & Boyd, Edinburgh, 1968).
Paton, L. A., *Studies in the Fairy Mythology of Arthurian Romance* (Burt Franklin, N.Y., 2nd edn, 1960).
Pearce, S. M., 'The Cornish Elements in the Arthurian Tradition', *Folklore*, volume 85 (Autumn 1974).
Perlesvaus, see *The High History*.
The Quest of the Holy Grail, trans. P. M. Matarasso (Penguin Books, 1969, paperback).
Rees, A. and B., *Celtic Heritage* (Thames & Hudson, London, 1973, paperback reprint).
Robert de Boron, *Le Roman dou L'Estoire dou Graal*, ed. W. A. Nitze (Librairie Honoré Champion, Paris, 1971 reprint) (the *Joseph* and the *Merlin*).
The Romance of Perceval in Prose, trans. D. Skeels (University of Washington Press, 1961) (the Didot *Perceval*).
Ross, Anne, *Pagan Celtic Britain* (Routledge & Kegan Paul, London, 1967).
——, *Everyday Life of the Pagan Celts* (Batsford, London, and Putnam, N.Y., 1970).
Sir Gawain and the Green Knight, ed. J. R. R. Tolkien and E. V. Gordon (Clarendon Press, Oxford, 1952 reprint).
——, trans. B. Stone (Penguin Books, 1959, paperback).
Stevens, John, *Medieval Romance* (Hutchinson, London, 1973).
The Tain, trans. T. Kinsella (Oxford University Press, 1970).
Treharne, R. F., *The Glastonbury Legends* (Cresset Press, London, 1967).
Vinaver, Eugene, *The Rise of Romance* (Clarendon Press, Oxford, 1971).

Wace and Layamon, *Arthurian Chronicles*, trans. E. Mason (Dent, London, and Dutton, N.Y., Everyman's Library, 1976 reprint).

Waite, A. E., *The Holy Grail* (University Books, N.Y., 1961 reprint).

Weston, Jessie L., *The Quest of the Holy Grail* (Frank Cass, London, 1964 reprint).

——, *From Ritual to Romance* (Doubleday Anchor Books, N.Y., 1957 reprint).

William of Malmesbury, *Chronicle of the Kings of England* (Bohn, London, 1847).

Wolfram von Eschenbach, *Parzival*, trans. H. M. Mustard and C. E. Passage (Random House, N.Y., 1961).

Index